Exploring the Universe

Richard Stimets

University of Massachusetts-Lowell

KENDALL/HUNT PUBLISHING COMPANY

4050 Westmark Drive   P.O. Box 1840   Dubuque, Iowa 52004-1840

Copyright © 2007 by Richard Stimets

ISBN: 978-0-7575-6070-5

Printed in the United States of America
10 9 8 7 6

Table of Contents

Preface

Since 1991, it has been my privilege to teach a one-semester course in astronomy entitled "Exploring the Universe" during both fall and spring semesters to a large class of mostly non-science majors at the University of Massachusetts, Lowell. It has been a challenge and also a mission to communicate both the basics and the rapidly unfolding discoveries and developments in astronomy and astrophysics to such a diverse audience. For most of these students, the lecture course and accompanying laboratory course will be their chief contact with university science courses, and it is important to achieve a basic understanding of the scientific method and an appreciation of the excitement of scientific research. These students will comprise the bulk of the educated voters on whose support rests the future health of scientific research in this country.

After seventeen years, I have decided to compile the essential ideas and insights which I have attempted to communicate into a book. This book is not only a comprehensive textbook, but it is also a study guide which provides a compact summary of the material as well as illustrative tables, figures and example problems for each chapter. The chapters arranged are in the order in which I teach the topics in my course. I have found over the years that this order and the rate of one chapter per week is good for the students.

My own training is as a physicist; and although this book does not use calculus, it does include several derivations based on fundamental laws of physics, e.g. Kepler's Third Law from Newton's Laws and the Law of Gravity, the Roche Limit from the Law of Gravity, and the Fate of the Earth–Moon System from the Conservation of Angular Momentum. These should appeal to both the non-science major who wishes to delve deeper than the presentation usually given in an introductory astronomy text as well as to the physics major who wants to see a real-world application of the basic laws which he/she has studied.

A number of years ago, I did some work in the fields of stellar spectroscopy and the solar-stellar connection, and this background is reflected in the inclusion of many stellar spectra and their analysis in my course. The spectra are uniformly presented on a logarithmic wavelength scale which facilitates:
(1) comparison of relative flux levels in different parts of the electromagnetic spectrum,
(2) comparison with blackbody spectra, and
(3) determination of velocity from the Doppler shift. The student who takes a little time to master the logarithmic wavelength scale will be amply rewarded by its simplicity and versatility.

Over the years, intriguing questions have arisen from both the perceptive students and my colleagues in the physics department. In order to answer these questions I have had to go beyond the material usually presented in introductory astronomy texts and have been helped by some good material available on the Internet. Many of these questions and answers are included in the book.

In order to keep the book to a modest length, I have often had to cover some topics briefly and omit others entirely where I felt that I could not do them justice with a brief discussion. Two examples are the important topics of global warming and the intriguing phenomenon of gravitational lensing. In such a vast field as astronomy, any brief book must be selective and supplementary.

The list of the nearly 300 brightest stars including contributions from the ultraviolet, visible, and infrared regions given in Appendix 8 is the culmination of several years of work, and has drawn from a number of sources including the Yale Bright Star Catalog, photometric data available from the Internet, and ultraviolet spectra from the OAO2, Copernicus, IUE and EURD satellites. In order to create a complete list, I have had to make a number of approximations and estimations. I think the current list is fairly accurate, but undoubtedly there will be some corrections and additions in future years. What is striking is how much this list of bright stars is dominated by the blue (B) and red (K and M) stars. It is a classic example of a bimodal distribution.

I would especially like to thank the following people for the help in preparing this book: Joe Sabella of Kendall-Hunt for encouraging me to write it in the first place; Jim Giddings for the preparation of the spectral figures, the star list and star name list; Dave Riccio for the preparation of the hand-drawn figures and the front cover; Sinang Chourb and Rebecca Rouse for the typing of the manuscript. It has been a great pleasure to work with all of them.

Richard Stimets
Lowell, Massachusetts
May 2007

Chapter 1: Introduction

Modern astronomy deals with the quantities which have an enormous range of values from the very small to the very large. Although it is possible to achieve a basic understanding of the essential concepts without calculus, it is not possible to do so without some mathematics and quantitative problem solving. In this introduction, we review some basic ideas which will prove to be useful in solving the problems and in interpreting and using the data presented in tables and graphs.

1.1 Calculating with Numbers Expressed in the Scientific Notation

Quantities varying over the large ranges encountered in astronomy are usually expressed in the following scientific notation:

$$\text{quantity} = \text{coefficient} \times 10^{exp} \times \text{units}$$

The fundamental physical constants and some important astronomical quantities are listed in Appendix A1.

Helpful Hint: If you use the following procedure, you are less likely to make errors while calculating with very small and large quantities:
1. Separate the combination of quantities into two groups: one group containing the coefficients and the another group containing the powers of 10.
2. Calculate a resulting coefficient by multiplying and dividing the coefficients on your calculator.
3. Calculate a resulting power of 10 by adding and subtracting the exponents.
4. Adjust the coefficient and the exponent to express the answer in the preferred format with one digit before the decimal point, e.g.

$$734.9 \times 10^{20} \text{ kg} \rightarrow 7.349 \times 10^{22} \text{ kg}$$

5. When using a formula containing a half-integer power, i.e. a square root, three-halves power, etc. adjust the coefficient and the power of 10 of the quantities, if necessary, so that the overall power of 10 is even.

Example Problem 1.1-1: Calculate the escape velocity of a white dwarf star which has the mass of the Sun and the radius of the Earth.

$$
\begin{aligned}
v_{esc} &= \sqrt{(2GM/R)} = \sqrt{(\ 2 \times 6.673 \times 10^{-11} \text{ m}^3/\text{kg/s}^2 \times 1.989 \times 10^{30} \text{ kg} \ / \ 6.378 \times 10^6 \text{ m})} \\
&= \sqrt{(2 \times 6.673 \times 1.989 / 6.378)} \times \sqrt{(10^{-11} \times 10^{30}/ 10^6 \text{ m}^2/ \text{ s}^2)} \\
&\quad -11 + 30 - 6 = 13: \underline{\text{odd}} \text{ so rewrite mass as } 19.89 \times 10^{29} \text{ kg} \\
&= \sqrt{(2 \times 6.673 \times 19.89 / 6.378)} \times \sqrt{(10^{-11} \times 10^{29}/ 10^6 \text{ m}^2/ \text{ s}^2)} \\
&\quad -11 + 29 - 6 = 12: \underline{\text{even}} \\
&\quad 12 / 2 = 6 \text{ (because of square root)}
\end{aligned}
$$

$$\boxed{v_{esc} = 6.451 \times 10^6 \text{ m/s.}}$$

1.2 Logarithmic Scales

It is often convenient to display the wide-ranging quantities encountered in astronomy on a graph on a logarithmic scale. The logarithm log x of a given number x is a number such that 10 raised to that power is x, i.e. $10^{\log x} = x$. The logarithms of the powers of 10 are integers, for example:

$$\log 1 = 0 \qquad 10^0 = 1 \qquad \log 0.1 = -1 \qquad 10^{-1} = 1 / 10^1 = 0.1$$
$$\log 10 = 1 \qquad 10^1 = 10 \qquad \log 0.01 = -2 \qquad 10^{-2} = 1 / 10^2 = 0.01$$
$$\log 100 = 2 \qquad 10^2 = 100$$

The logarithms of other numbers are not integers but are sometimes <u>close to round decimal fractions.</u> For example, enter the number 2 in your calculator and execute the log x function. You will get:

$$\log 2 = 0.301029996 \cong 0.3$$

Similarly, if you enter 0.3 and execute the 10^x function, you will get

$$10^{0.3} = 1.995262315 \cong 2$$

For the purpose of a quick mental calculation, we may regard log 2 as being 0.3 and $10^{0.3}$ as being 2. The error introduced by this approximation is less than 1% and can often be neglected.

Table 1.2-1 shows the decimal fraction log x on a scale from 0 to 1 and the corresponding approximate number x from 1 to 10. An interval comprising a change of 1 unit on the log x scale, and factor of 10 on the x scale is termed a <u>decade</u> (it should not be confused with its meaning as a unit of time equal to 10 years). A change of 0.5 is half a decade, a change of 0.3 is three-tenths of a decade, etc.

Table 1.2-1 Fractions of a Decade and Corresponding Factors

log x	0	0.1	0.2	0.3	0.4	0.5	0.6	0.7	0.8	0.9	1.0
x	1.0	1.25	1.58	2.0	2.5	3.16	4.0	5.0	6.3	8.0	10.0

Factors for the larger logarithmic intervals can be easily derived by using the correspondence between the sum of logarithms and the product of their respective factors

$$\log x_1 + \log x_2 = \log (x_1 \times x_2) \qquad \text{(Eq 1.2-1)}$$

Use the following procedure:
1. Break up a larger logarithmic value into a sum of smaller values whose corresponding factors are known.
2. Multiply these factors to get the overall factor.

<u>Example Problem 1.2-2:</u> Find the corresponding factors for the logarithmic values 1.7 and 2.5.

log values:	$1.0 + 0.7 = 1.7$	$2.0 + 0.5 = 2.5$
factors:	$10 \times 5 = 50$	$100 \times 3.16 = 316$

Two important uses of logarithmic scales in this book are the logarithmic wavelength scale and the stellar magnitude scale.

If logarithmic scales are used on both the abscissa and the ordinate of a graph, one has a log–log graph. Such graphs have the useful property that they display a power law as a linear relation:

$$y = ax^n$$
$$\log y = \log a + n \log x \qquad \text{(Eq. 1.2-2)}$$

Thus, the logarithm of the constant a is the intercept and the exponent n is the slope on a graph of log y vs. log x. The well-known procedure of linear regression (see Section 1.5) can be used to derive a best-fit straight line from the data. Although non-linear procedures, such as quadratic regression, cubic regression, etc. exist, sometimes, in nature, the exponents of power laws are not integers. In addition, human observers can easily see a straight-line pattern of points on a graph but find it very difficult to distinguish a quadratic from a cubic curve, etc.

1.3 Percent Error

The concept of percent error is important, both in comparing a theoretical or predicted value of a quantity with a measured value and in comparing a student's measured value with an accepted measured value. If x is the theoretical, predicted, or student's measured value and x_{acc} is the accepted measured value

$$\% \text{ Error} = (x - x_{acc}) / x_{acc} \times 100\% = (x / x_{acc} - 1) \times 100\% \qquad \text{(Eq. 1.3-1)}$$

The sign of the error is either positive or negative according to whether the test value is greater or less than the accepted value.

<u>Example Problem 1.3-1:</u> Bode's law predicts the distance of Mars from the Sun to be 1.6 AU whereas the accepted value is 1.524 AU. Find the percent error.

$$\% \text{ Error} = (1.6 - 1.524) / 1.524 \times 100\% = 5.0\%$$

If we are comparing logarithmic values, the following relation must be used:

$$\% \text{ Error} = [10^{(\log x - \log x_{acc})} - 1] \times 100\% \qquad \text{(Eq. 1.3-2)}$$

Note that because $10^{\log x} = x$ and $10^{\log x_{acc}} = x_{acc}$, this relation is equivalent to the second form above.

Example Problem 1.3-2: A linear regression fit to the distances of the planets from the Sun predicts that the distance of the Earth in AU should be given by log d = .027, whereas the accepted value is d = 1 AU, log d = 0.0. Find the % error.

$$\% \text{ Error} = [10^{(.027-.00)} - 1] \times 100\%$$
$$\% \text{ Error} = [1.064 - 1] \times 100\%$$

$$\boxed{\% \text{Error} = 6.4\%.}$$

Caution: Note that if we had mistakenly tried to substitute the logarithmic values in the first formula for % error given above, we would have been dividing by zero and gotten infinity, which is absurd. In comparing original values, we take ratios; in comparing logarithmic values, we take differences.

1.4 Statistics

Astronomy deals with large numbers of objects in a class such as stars, star clusters, galaxies, etc. In characterizing their properties such as positions, distances, masses, luminosities, etc., one often makes use of statistical quantities and methods.

The Mean
Assume that we have a class of N objects, such as stars, and that we are considering the values x, of a property associated with them, such as position. The mean value x_{ave} (read as 'x average') of the quantity x for the class is defined as:

$$x_{ave} = (x_1 + x_2 + \dots x_N)/N$$

$$x_{ave} = (\Sigma x_i)/N \qquad \text{(Eq. 1.4-1)}$$

where the upper case Greek letter sigma stands for a sum over the index i of the objects from 1 to N.

The Standard Deviation
The mean value by itself is not sufficient to characterize the entire set of values, however. Two sets of values might have the same mean, but, in one case, the range of values about the mean might be quite small, whereas in another case, it was quite large. A measure of spread of the values about the mean is essential. The two quantities which measure this spread are the variance (Eq. 1.4-2) and its square root, the standard deviation (Eq. 1.4-3).

$$\sigma^2(x) = [(x-x_{ave})^2 + (x_2 - x_{ave})^2 + \dots (x_N - x_{ave})^2]/N$$
$$\sigma^2(x) = (1/N) \Sigma(x_i - x_{ave})^2 \qquad \text{(Eq. 1.4-2)}$$
$$\sigma(x) = \sqrt{\sigma^2(x)} \qquad \text{(Eq. 1.4-3)}$$

Note that the standard deviation has the same units as the mean, whereas the variance has these units squared. For example, if x is in meters, x_{ave} and $\sigma(x)$ are also in meters, whereas $\sigma^2(x)$ is in meters squared. Sometimes the variance and standard deviation are computed with a factor N-1 instead of N, but there is little difference for N >10.

The Coefficient of Variation
Another quantity, derived from the mean and the standard deviation, measures the <u>relative spread</u> of the values about the mean. This is the coefficient of variation c(x), defined as the quotient of the standard deviation and the mean:

$$c(x) = \sigma(x)/x_{ave} \qquad\qquad \text{(Eq. 1.4-4)}$$

Unlike the mean and the standard deviation, which usually have units, the coefficient of variation is a dimensionless quantity and can be expressed as a fraction or a percentage.

<u>Example Problem 1.4-1</u>: A team of basketball players has the following heights: 1.9 m, 1.95 m, 2.0 m, 2.05 m, and 2.1 m. Find the mean height h_{ave}, the standard deviation $\sigma(h)$, and the coefficient of variation c(h).

$$h_{ave} = (1.9 + 1.95 + 2.0 + 2.05 + 2.1)/5 \text{ m}$$

$$h_{ave} = 2.0 \text{ m}$$

$$\sigma(h) = .079 \text{ m (from calculator)}$$

$$c(h) = .079 \text{ m}/2.0 \text{ m}$$

$$c(h) = .0395 = 3.95\%$$

<u>Caution:</u> The coefficient of variation is not always meaningful, particularly for quantities which have an arbitrary zero point to the scale, such as time, position, or temperature. For example, let us assume that the mean daily temperature for December in a certain city is 0°C (32°F) and that the standard deviation is 5°C (9°F). If we compute the coefficient of variation on the Fahrenheit scale, we get 9°F / 32°F = .281 = 28.1%, whereas on the Celsius scale, we get 3°C / 0°C = ∞, which is, of course, absurd! The problem is the arbitrary zero point of these two temperature scales. The scale which we should use in order to get a meaningful result is the Kelvin scale. Then the mean temperature is 273 K, the standard deviation is 5 K, and the coefficient of variation is 5 K/273 K = .0183 = 1.83%.

Distribution Functions
It is often useful and revealing to display the distribution of values of a quantity as a graph. For a small or moderate number of objects in a class, the distribution can be displayed as a <u>histogram</u>, which shows the fraction of objects having values within various ranges. Figure 1.4-1 is such a histogram which shows the fraction of planets in the solar system

having the logarithms of their radii (relative to Earth) between −0.5 and 0.0, 0.0 and 0.5, 0.5 and 1.0, etc.

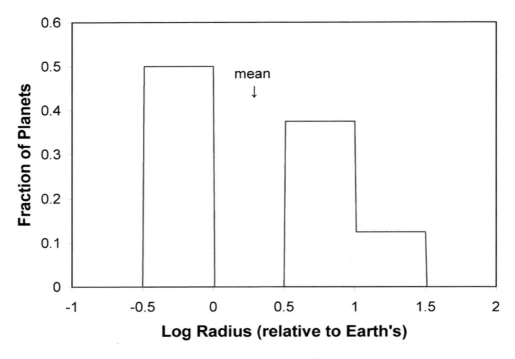

Fig. 1.4-1. Distribution of Planetary Radii for the Solar System.

As the number of objects in the class becomes large, the distribution of values of an associated quantity may be approximated by a continuous curve. Two such distribution functions, namely the normal distribution function and the cosine distribution function, are shown in Figs. 1.4-2 and 1.4-3, respectively. In Fig. 1.4-2, the units of the abscissa are standard deviations, which have the same units as the quantity being measured, e.g. meters for distance, height, or radius and kilograms for mass, etc. The units of the ordinate are probability per standard deviation. The probability of the abscissa lying within a certain range is equal to the area under the curve and can be estimated as the product of the range of the abscissa multiplied by the average value of ordinate over the range. In Fig. 1.4-3, the units of the abscissa are radians (or equivalently degrees) and the units of the ordinate are probability per radian.

Example Problem 1.4-2: Estimate the probability of the value of the abscissa in a normal distribution lying between −1.5σ and +1.5σ.

$$\text{range of abscissa } 1.5\sigma - (-1.5\sigma) = 3.0\sigma$$
$$\text{average ordinate} \approx .27/\sigma$$

$$\boxed{\text{probability} \approx 3.0\sigma \times .27/\sigma = .81}$$

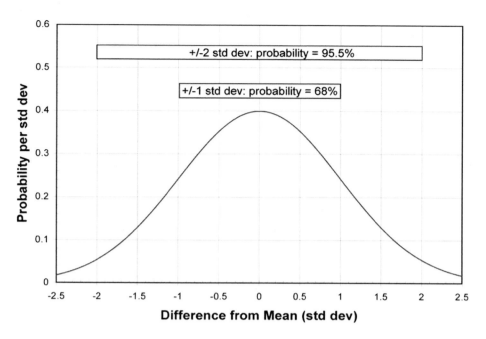

Fig. 1.4-2. The Normal Distribution Function.

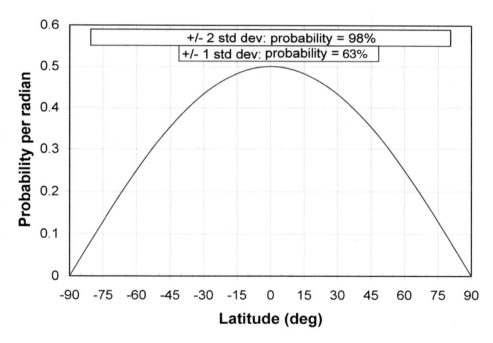

Fig. 1.4-3. The Cosine Distribution Function.

The normal distribution function can be expressed mathematically as:

$$p(x) = (1/\sqrt{(2\pi)}) \exp(-x^2/2) \qquad \text{(Eq. 1.4-5)}$$

There is no corresponding simple formula for the area under the curve but the values are tabulated in standard mathematical handbooks.

Here is an important example of the use of the cosine distribution function in astronomy:
1. Assume that the stars in the sky are uniformly distributed over a sphere.
2. In some chosen coordinate system, what is the distribution of the values of latitude of the stars?

It obeys the cosine distribution function, which has its maximum at the equator and goes to zero at the poles. The abscissa is marked off in degrees for convenience, but its natural units are radians and the standard deviation is .683667390 radians = 39.1712560°.

The cosine distribution function can be expressed mathematically as:

$$p(\theta) = (1/2) \cos\theta \qquad\qquad \text{(Eq. 1.4-6)}$$

In this case, the area under the curve between two latitudes θ_1 and θ_2 may also be expressed by a simple formula:

$$\text{probability} = \frac{1}{2} (\sin\theta_2 - \sin\theta_1) \qquad\qquad \text{(Eq. 1.4-7)}$$

Example Problem 1.4-3: Estimate the probability that, in a population of uniformly distributed stars, a given star has a latitude between 0° and 30°.

$$\boxed{\text{probability} = (\tfrac{1}{2})\,(\sin 30° - \sin 0°) = \tfrac{1}{4}}$$

A knowledge of the cosine distribution function allows one to determine quantitatively whether a given class of stars is uniformly distributed or not by comparing its standard deviation of latitude with the uniform-distribution value of 39.1712650°. As we shall see later on, some stellar populations are uniformly distributed and some are not, being more concentrated toward the equator, in particular, the galactic equator.

Unimodal and Bimodal Distributions
The normal distribution function and the cosine distribution function are both examples of a <u>unimodal</u> distribution function, which has a single peak and tapers to zero at large deviations from the mean value. They are also <u>symmetric</u>: the peak value of the ordinate is located at the mean value of the abscissa so that the most probable value of the abscissa is the mean value. Values deviating by an equal amount positively and negatively from the mean have equal probability.

Some distributions are unimodal but <u>asymmetric</u>. An example from astronomy would be the distribution of stellar masses relative to the Sun's mass. In this case, the most probable value lies somewhat below the mean value and there is a long tail extending up to many times the mean value. The amount of asymmetry is a function of the scale used in plotting and the stellar mass distribution would look more symmetric on a logarithmic scale of mass, but would still be skewed to the high-mass side.

Other distributions are not unimodal but <u>bimodal</u>, having two distinct peaks. The distribution of planetary radii shown in Fig. 1.4-1 is one such distribution, in which the total population divides into a group of small-radius planets, i.e. the terrestrial planets, and a group of large-radius planets, i.e. the Jovian planets. The mean value on the logarithmic scale is .313 (i.e. a radius a little larger than twice the Earth's). It is indicated on the graph but has little meaning since no planet has anywhere near this value of the radius.

It is often possible and desirable with bimodal distributions to separate them into two unimodal distributions, each of which can be characterized by its own mean and standard deviation. The occurrence of a bimodal distribution usually indicates that <u>different physical processes</u> were involved in producing the two groups of objects.

1.5 Curve Fitting: Linear Regression

The most commonly used curve-fitting technique is linear regression, which is available on most scientific calculators. Given two sets of points $[x_i]$ and $[y_i]$ for $i = 1:N$, the procedure finds a line

$$y = mx + b \qquad\qquad \text{(Eq. 1.5-1)}$$

with the slope m and intercept b such that the sum of the squares of the differences between the calculated and given values of y is a minimum. As noted in Section 1.2, if logarithmic scales are used, this straight line represents a power law relation between x and y in which the slope is the exponent and the intercept is the coefficient.

As an example of linear regression, consider the data for planetary distances (in AU) in the solar system listed in Table 1.5-1.

Table 1.5-1 Planetary Distances in the Solar System

Position #	Planet	Distance (AU)	log Distance
1	Mercury	0.387	–0.412
2	Venus	0.773	–0.141
3	Earth	1.000	0.000
4	Mars	1.524	0.183
5	(Ceres)	2.8	0.447
6	Jupiter	5.203	0.716
7	Saturn	9.539	0.980
8	Uranus	19.19	1.283
9	Neptune	30.06	1.478

Figure 1.5-1 shows a plot of log distance versus position number for nine objects, namely the eight planets and the largest asteroid Ceres, together with the linear regression fit (the fit is much worse if Ceres is excluded). The slope of 0.239 means that, on average, each object is $10^{0.239} = 1.734$ times further from the Sun than the previous one. Although the fit is not perfect, the correlation coefficient of 0.997 means that it is very good. The distances predicted by this linear regression fit are different, although similar, to those predicted by Bode's Law (see Section 6.2).

Caution: Although the example given here clearly illustrates the procedure of curve fitting with linear regression, the resultant linear relation is not based on any clearly evident underlying physical principles. However, there are other cases in astronomy, e.g. the mass–luminosity relation along the main sequence, and the period–luminosity relation for Cepheid variables, where a linear fit has a firm basis in the underlying physics.

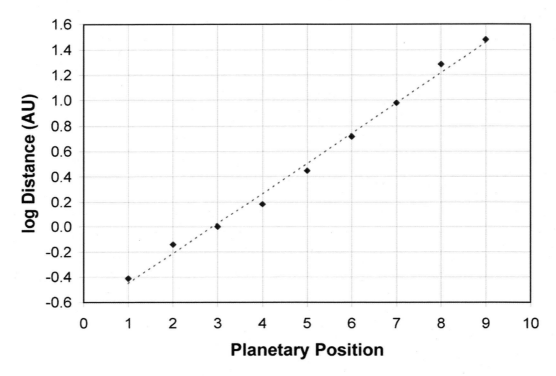

Fig. 1.5-1

Questions

1.1. Why is the scientific notation used to express quantities in astronomy?
1.2. Explain the procedure for finding the corresponding factor for a larger logarithmic interval which is a sum of two smaller ones.
1.3. What kind of relation is displayed as a linear relation on a log–log graph?
1.4. Give the two different formulas for calculating percent error using
(a) original values and
(b) their logarithms.
1.5. Why is the mean insufficient to characterize a set of values?
1.6. The velocities in a set of values have units of km/s. What are the units of the mean, the standard deviation and the coefficient of variation of the set?
1.7. Give two examples of distribution functions.
1.8. How can we tell quantitatively whether a given class of stars is uniformly distributed or not?
1.9. Explain the difference between unimodal and bimodal distributions.
1.10. If we plot a set of values on a log–log graph and perform a linear–regression fit, what is the meaning of the slope m?

Problems

1-1. Find the corresponding factors for the logarithmic values 1.6 and 1.9.
1-2. Suppose the points on a log–log plot can be fit with a linear relation of the form $\log y = .3 + 1.5 \times \log x$. Find the corresponding power law.
1-3. Suppose the accepted value of a stellar distance is $d_{acc} = 20$ light years and a student gets a value $d = 25$ light years. Using Eq. 1.3-1, what is the percent error? What are the logarithms of these values? Using Eq. 1.3-2, what is the percent error?
1-4. Calculate the mean and standard deviation of the galactic latitude for the O stars listed in Appendix A8. Are the stars uniformly distributed or concentrated toward the galactic equator?

Chapter 2: "Fixed Stars" and Moving Planets

Early astronomy was mainly concerned with trying to explain the motion of the planets. Because of immense distances of the stars, their motion is much slower and was not noticed until the late eighteenth century. All early models of planetary motion were based on circles. The geocentric Ptolemaic model achieved fairly good accuracy, but at the cost of an elaborate, inelegant scheme involving epicycles and off-center orbits. The heliocentric Copernican model achieved an elegant simplicity, but at the cost of accuracy. Finally, when Kepler combined the idea of elliptical orbits with the heliocentric model, he produced a description of the solar system which was both elegant and accurate.

2.1 Astronomical Coordinate Systems

Astronomers, like geographers, need a system of longitude and latitude to specify the location in the sky of planets, stars and other celestial objects. Two systems are in common use.

Celestial Coordinates

Celestial coordinates are based on a projection of the Earth's coordinate system onto the sky. Longitude is referred to as <u>right ascension</u> and latitude as <u>declination</u>. The big difference is that right ascension is measured in hours, minutes and seconds of time rather than degrees of angle, with the equivalence that 24 hours equals 360°. Conversion of right ascension and declination to decimal degrees is accomplished in a straightforward manner by the use of the following table.

Table 2.1-1. Conversion of Right Ascension and Declination to Decimal Degrees

Right Ascension (RA)	Declination (DEC)
1 hour (h) = 15°	1 degree (°) = 1 degree
1 minute (m) = .25°	1 minute (′) = .016667°
1 second (s) = .004167°	1 second (″) = .000278°

<u>Example Problem 2.1-1:</u> The coordinates of the star Polaris for the year 2000 are: RA = 2h 31m 50.5 s, DEC = 89° 75′ 51″. Convert these values to decimal degrees.

$$RA \rightarrow 2 \times 15° + 31 \times .25° + 50.5 \times .004167° = 37.960434°$$

$$DEC \rightarrow 89° + 75 \times .016667° + 51 \times .000278° = 89.264183°$$

Because of the precession of the Earth's axis, the celestial coordinates of stars change a little bit each year and are specified in star atlases for a round-number year such as 2000 or 2050.

Galactic Coordinates

Galactic coordinates are based on a galactic equator, which lies in the plane of our galactic disk. The zero point of longitude lies toward the direction of the galactic center. Galactic coordinates are already expressed in decimal degrees, so no conversion is necessary. They are more useful than celestial coordinates for telling how a class of stars is located, oriented and distributed with respect to the galactic disk. The positions of the stars listed in Appendix A8 are expressed in galactic coordinates.

2.2 Size and Angular Size

In astronomy, when viewing an object from a large distance, we measure its angular size. Its true size may be calculated if the distance is known. If an object having a diameter D is viewed from a distance d and has an angular size α, the relation is:

$$D = \alpha d \qquad\qquad \text{(Eq. 2.2-1)}$$

where D and d are in the same units, e.g. kilometers, and α is in radians (Note: a radian is dimensionless and does not change the units on the right hand side). Figure 2.2-1 depicts the relation geometrically, and clearly shows the inverse relation between α and d for a given value of D.

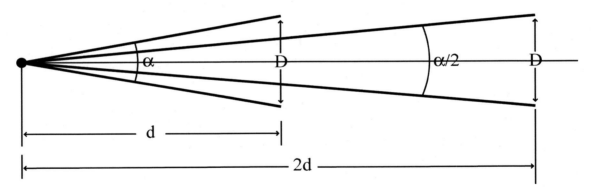

Fig. 2.2-1 Size and Angular Size

If α and d are in other units, e.g. arc seconds and astronomical units (AU), they must first be converted to radians and kilometers using the conversions 1 arc sec = 4.848×10^{-6} radians and 1 AU = 1.50×10^{8} km.

<u>Example Problem 2.2-1:</u> The planet Jupiter has an angular diameter of 46.8 arc seconds. Given that Jupiter's distance from the Sun is 5.2 AU at opposition when the alignment is Sun–Earth–Jupiter, how far is Jupiter from Earth? What is Jupiter's diameter in kilometers?

$$d = 5.2 \text{ AU} - 1.0 \text{ AU} = 4.2 \text{ AU} = 4.2 \times 1.50 \times 10^8 \text{ km}$$

$$\boxed{d = 6.3 \times 10^8 \text{ km}}$$

$$\alpha = 46.8 \times 4.848 \times 10^{-6} \text{ rad} = 2.27 \times 10^{-4} \text{ rad}$$
$$D = \alpha d = 2.27 \times 10^{-4} \times 6.3 \times 10^8 \text{ km}$$
$$D = 2.27 \times 6.3 \times 10^4 \text{ km}$$

$$\boxed{D = 1.43 \times 10^5 \text{ km}}$$

2.3 Velocity and Angular Velocity

The treatment of velocity and angular velocity is similar to that for size and angular size. Let us assume that in a time interval Δt, a celestial object travels a distance Δx so that its velocity is $v = \Delta x/\Delta t$ and $\Delta x = v\Delta t$. In angular measure, it travels an angular distance $\Delta\theta$ so that its angular velocity is $\omega = \Delta\theta/\Delta t$ and $\Delta\theta = \omega\Delta t$. Referring back to the previous section, we see that Δx plays the role D and $\Delta\theta$ plays the role of α so that

$$\Delta x = \Delta\theta d \qquad\qquad \text{(Eq. 2.3-1)}$$

and dividing both sides by Δt we have

$$v = \omega d \qquad\qquad \text{(Eq. 2.3-2)}$$

Figure 2.3-1 depicts this relation geometrically and clearly shows the inverse relation between ω and d for a given value of v.

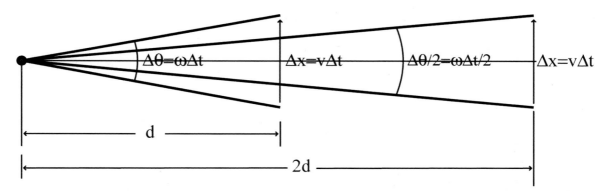

Fig. 2.3-1. Velocity and Angular Velocity.

In using Eq. 2.3-2, v has the units d per second and ω is in radians per second. Otherwise ω must first be converted to rad/sec.

Example Problem: The Moon moves through the sky at an average angular velocity of 12° per day. Its average distance from the Earth is 3.84×10^5 km. What is its average velocity?

$$12° = .2094 \text{ rad}$$
$$1 \text{ day} = 86400 \text{ s}$$
$$\omega = .2094 / 86400 \text{ rad/s} = 2.424 \times 10^{-6} \text{ rad/s}$$
$$v = \omega d = 2.424 \times 10^{-6} \text{ rad/s} \times 3.84 \times 10^5 \text{ km}$$
$$v = 2.424 \times 3.84 \times 10^{-1} \text{ km/s}$$

$$\boxed{v = .93 \text{ km/s}}$$

In their view of the heavens, the ancient astronomers reasoned that relative distance of the planets varied inversely as relative angular velocity. Essentially, they were using Eq. 2.3-2 and assuming that v was approximately the same for different planets. We now know from Kepler's Third Law (see Section 2.7) that v falls off inversely as the square root of the distance so that the outer planets are moving slower than the inner ones. For example, Saturn's average angular velocity is about .4 times that of Jupiter. Assuming equal velocities, one would conclude that Saturn is 2.5 times as far away as Jupiter whereas the true value is approximately 2 times. However, the error introduced by the assumption of equal velocity is not enough to change the <u>order of distance</u> of the planets.

2.4 The Earth: Fixed or Moving?

How do we know that the Earth is moving? The idea of a fixed Earth was so entrenched that, even after Copernicus put forth his heliocentric theory in 1543, brilliant astronomers, such as Tycho Brahe, refused to accept the idea of a moving Earth. In Tycho's hybrid model, the other planets orbited the Sun, but the Earth was still fixed and the Moon and Sun orbited the Earth. Because such a model yields relative positions of the Sun, Moon, and planets, which are identical to those of the fixed-Sun heliocentric model, there is no way from observations of solar-system objects alone to distinguish them.

There are two phenomena which definitely prove that it is the Earth which is moving and not the Sun:

1. Parallax: As the Earth makes its annual journey around the Sun, our viewing angle of the nearby stars against the more distant ones changes slightly, and they trace out cyclic paths in the sky. However, even for the nearest bright star, <u>Alpha Centauri</u>, the total excursion from the average position is less that one second of arc. Both a telescope and photography are necessary to observe it, and this observation was first accomplished by Friedrich Bessel in 1838, who measured the parallax of the nearby star 61 Cygni, as 0.314 arc seconds.

2. Aberration of Starlight: Because the Earth is moving, the apparent direction from which the light of a star appears to originate is slightly shifted. The effect is similar to that encountered when driving in a snowstorm with very light wind, in which the snow is really coming straight down but appears to be coming from a direction somewhat above the horizontal due to the motion of the car. The maximum angular shift of position $\Delta\theta$ is given by:

$$\Delta\theta \text{ (radians)} = v/c \qquad \text{(Eq. 2.4-1)}$$

which for the Earth is 30 km/s /300000 km/s = 10^{-4} radians = 20.6 arc seconds. A telescope is necessary to observe the effect, and the first observation was made by James Bradley in 1727 of the star Gamma Draconis.

Since Tycho Brahe died in 1601, more than a century before there was any definitive evidence of the Earth's motion, his reluctance to accept the idea of a moving Earth is understandable.

2.5 Kepler's First Law: The Shape of the Orbit

Kepler's first law states that the orbits of the planets are ellipses with the Sun at one focus. Earlier models of planetary orbits had been based on centered circles, off-center circles and circles upon circles (epicycles), but Kepler was the first astronomer to use ellipses. A representative ellipse with its key points and dimensions is shown in Fig. 2.5-1.

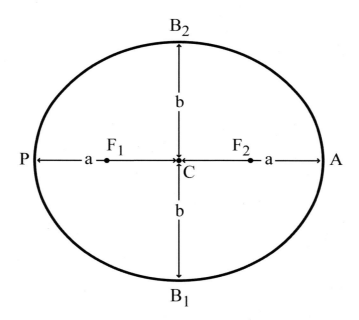

Fig. 2.5-1. Key Points and Dimensions of an Ellipse.

The eccentricity e is a measure of its non-circularity and may be expressed as the ratio of the center-to-focus distance CF_1 divided by the semimajor axis a (= CA = CP);

$$e = CF_1 / CA = CF_1 / CP \qquad \text{(Eq. 2.5-1)}$$

or equivalently by the ratio of the difference of the aphelion distance r_A and the perihelion distance r_P to their sum:

$$e = (r_A - r_P) / (r_A + r_P) \qquad \text{(Eq. 2.5-2)}$$

The relation between the major axis a, the minor axis b and the eccentricity e is:

$$b/a = \sqrt{ (1 - e^2) } \qquad \text{(Eq. 2.5-3)}$$

In order for the axial ratio b/a to differ noticeably from 1, the eccentricity e must be fairly large. Values of e and b/a are shown for several elliptical orbits in Table 2.5-1 below.

Table 2.5-1: Eccentricity e and Axial Ratio b/a for Some Elliptical Orbits

Planetary Orbit	Eccentricity (e)	Axial Ratio (b/a)
Earth	.017	.99986
Mars	.093	.99567
Fig. 2.5-1	.44	.8980

Observing Elliptical Orbits from Another Orbit
The problem which Kepler faced was to construct the planetary orbits from observations made from Earth, which is moving along its own elliptical orbit. This was not an easy task! What allowed him to do it successfully were the values of e and b/a for the Earth and Mars given in Table 2.5-1. For the Earth, e is small enough that its orbit could be approximated by an off-center circle, whereas for Mars, the difference of the orbit from an off-center circle was noticeable, with the data of Tycho Brahe available to Kepler.

Kepler's procedure was as follows:
1. Construct the orbit of the Earth (an off-center circle).
2. Assume that the Earth moves along its orbit according to the law of areas (Kepler's second law).
3. Take the angular positions of Mars from two observations 687 days apart (one full revolution of Mars). (Mars must have been at the same position in space when these observations were made but the Earth was at two different positions.)
4. From the baseline connecting the two positions of the Earth together with the two angular positions of Mars, construct a triangle to determine the position of Mars.
5. Repeat the procedure for other pairs of observations 687 days apart to construct the full orbit of Mars.

The orbit of Mars so constructed could not be accurately fit by a circle, an off-center circle, or an oval, but was very nicely fit by an ellipse with the Sun at one focus. In addition, the speed of Mars along its orbit varied in accordance with the law of areas assumed in Step 2 of the procedure. If Earth's orbit had been significantly more eccentric and/or Mars' orbit had been significantly less so, Kepler would not have been able to deduce his first law.

2.6 Kepler's Second Law: Motion along the Orbit

Kepler's second law states that a planet sweeps out equal areas in equal times as it moves along its orbit. This equality of areas is depicted in Fig. 2.6-1 at three points on the orbit: aphelion A, perihelion P, and a general point. In order to maintain equality of the shaded areas in the same time interval Δt, the planet must move faster when it is closer to the Sun and slower when it is farther away.

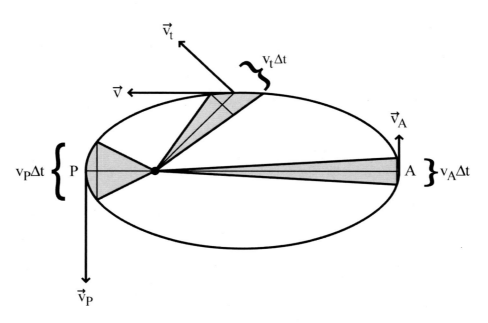

Fig. 2.6-1. Kepler's Law of Equal Areas.

Take particular note of the general point. In contrast to points A and P, where the velocity is all tangential, i.e. perpendicular to the radial direction, the velocity at the general point has both radial and tangential components and it is only the <u>tangential</u> component that contributes to the area swept out. Kepler's second law may be summarized by the equation:

$$v_t r = \text{constant} \qquad \qquad \text{(Eq. 2.6-1)}$$

where v_t is the tangential component of the velocity.

Kepler's second law is one example of a general conservation law in physics known as the <u>conservation of angular momentum</u>. It states that, because the attractive gravitational force from the Sun is completely radial, the planet maintains constant angular momentum as it moves along its orbit.

2.7 Kepler's Third Law: Comparison of Different Orbits

Kepler's third law states that the square of the orbital period P of a planet is proportional to the cube of its semimajor axis a:

$$P^2 = ka^3 \qquad \text{(Eq. 2.7-1)}$$

If P is measured in years and a in astronomical units, then k = 1 and we have:

$$P^2 = a^3 \text{ (P in years, a in AU)} \qquad \text{(Eq. 2.7-2)}$$

Unlike the first two laws, which dealt with the shape of, and motion along a single orbit, respectively, the third law compares different orbits. In Section 3.4, we will derive it from Newton's laws of motion and the universal law of gravitation and determine k in terms of other quantities. A consequence of Kepler's third law is the dependence of orbital speed on distance from the force center. For a circular orbit, a = the radius r and we have

$$P^2 = kr^3 \qquad \text{(Eq. 2.7-3)}$$

Using the relation that the speed along a circular orbit is equal to the circumference divided by the period, i.e. $v = 2\pi r/P$, we get:

$$v = 2\pi /\sqrt{k}r \qquad \text{(Eq. 2.7-4)}$$

Thus, the speed falls off inversely as the square root of the radius of the orbit.

2.8 Distance from Parallax

Astronomers had long realized that the distance of a celestial object could be determined if its parallax, i.e. its shift in angular position with a change in viewing position, could be measured. The geometry is shown in Fig. 2.8-1.

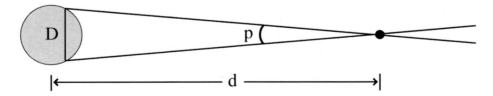

Fig. 2.8-1. Geometry of Parallax Measurement.

The distance d is calculated from the baseline D and the parallax angle p (in radians) by the relation:

$$d = D / p \qquad \text{(Eq. 2.8-1)}$$

Note that the geometry of Fig. 2.8-1 is similar to that of Fig. 2.1-1 but in reverse: The "diameter" is the baseline and is at the viewing position, not at the object position.

The difficulty is that, in viewing other solar system objects from the Earth, the baseline is restricted to be no longer than the diameter of the Earth and, for all objects other than the Moon, the parallax angle is at most a few arc seconds. This amount was immeasurable at the time of Tycho Brahe. Copernicus, Brahe, and Kepler all had a good idea of the relative distances of the planets but not their absolute values, although they suspected that it was probably at least tens of millions of miles.

By the 1670s telescopes had improved and accompanying hardware such as micrometer eyepieces with adjustable cross-hairs was available, so that angular separations in the arc-second range could be measured. During an opposition of Mars in the fall of 1672 (when Mars was also at perihelion so that its distance to Earth was a minimum), both Giovanni Cassini in Paris and John Flamsteed in Derby, England were able to measure the parallax of Mars, calculate its distance, and hence determine a scale for the entire solar system.

Cassini accomplished his measurement with the help of a colleague on Cayenne Island off South America. During the opposition, the two men made <u>simultaneous</u> measurements from <u>two different locations</u> of the angular separation between Mars and several background stars. When the colleague returned to Paris a year later, they compared values and derived a parallax. Flamsteed made all his measurements from <u>a single location</u> but at <u>two different times</u>, in evening and early morning, so that the baseline was created by the rotation of the Earth. Flamsteed wisely chose the night of October 6, 1672, when Mars was at the turnaround point of its retrograde motion, so that its own angular displacement during the time interval was minimal.

Based on their value of the distance of Mars, both Cassini and Flamsteed calculated the value of the astronomical unit as 87 million miles – about 6% lower than the currently accepted value.

2.9 Proper Motion of Stars

Until the eighteenth century astronomers had believed that the stars were fixed in position. In the late 1700s, Edmund Halley found differences in his current positions of the bright star Sirius from those in Ptolemy's catalog and realized that Sirius, and presumably other stars as well, were slowly moving across the sky. This motion is called proper motion and it is denoted by the symbol μ, usually measured in arc seconds per year. The angular displacement $\Delta\theta$ during a time interval Δt given by:

$$\Delta\theta = \mu\Delta t \qquad \text{(Eq. 2.9-1)}$$

We now know that Sirius moves at a rate of 1.328 arc seconds per year.

The star with the largest proper motion is a nearby red dwarf star named Barnard's Star in the constellation Ophiuchus, which is 5.91 light years away and moving through the sky at a rate of 10.31 arc seconds per year.

Example Problem 2.9-1: What is the angular displacement of Barnard's Star during a time interval of 100 years?

$$\Delta\theta = \mu\Delta t$$
$$\Delta\theta = 10.31 \text{ arc seconds/ yr} \times 100 \text{ yrs}$$

$$\boxed{\Delta\theta = 1031 \text{ arc sec.}}$$

This angular displacement is more than half the diameter of the full moon. The motion of Barnard's star would be clearly noticeable during a person's lifetime if it were a bright star, but, like all red dwarfs, it is too faint to be seen with the naked eye.

Questions

2-1. What are right ascension and declination?

2-2. In what way are galactic coordinates more useful than celestial coordinates?

2-3. What are the units of angular size?

2-4. What are the units of angular velocity?

2-5. What assumption did ancient astronomers make in using angular velocity to gauge distance?

2-6. What two phenomena prove that it's the Earth which is moving and not the Sun?

2-7. What is the eccentricity e for a circular orbit?

2-8. What allowed Kepler to deduce the ellipticity of Mars' orbit while he was moving on another elliptical orbit, namely the Earth's?

2-9. At a general point along the orbit, which component of the velocity contributes to the area swept out?

2-10. For a circular orbit, how does the orbital speed depend on the radius of the orbit?

2-11. When and how did astronomers first determine a scale for the entire universe?

2-12. Why is the large proper motion of Barnard's star not noticeable during a person's lifetime?

Problems

2-1. The coordinates of the star Sirius for the year 2000 are: RA = 6 h 45 m 8.9 s, DEC = −16° 42′58″. Convert these values to decimal degrees.

2-2. Jupiter moves through the sky at an average angular velocity of .083° per day. Its average distance from the Earth is 5.2 AU. What is its average velocity in km/s?

2-3. An elliptical orbit has an eccentricity of .6. What is the axial ratio b/a?

2-4. Given that the average orbital speed of the Earth (a = 1 AU) is 30 km/s, what is the average orbital speed of Jupiter (a = 5.2 AU)? Compare the answer with the answer for Problem 2-2.

2-5. How long would it take Barnard's star to move 5° across the sky?

Chapter 3: Gravity and Motion

The success of Kepler's empirical laws in describing the motion of the planets motivated Isaac Newton to come up with a set of more basic physical laws from which Kepler's laws could be derived. Newton's guiding principle was that these physical laws should be <u>universal</u> and apply equally to objects near the Earth and objects in the heavens.

3.1 Newton's First Law: Motion (or not) Without a Force

Newton's first law states that a body at rest or <u>in motion at constant velocity</u> tends to remain in that state unless acted upon by an outside force. The underlined phrase is the key difference between Newton's idea of motion and the earlier idea espoused by Aristotle. In the Aristotelean view, bodies near the Earth remained at rest unless acted upon by a force, whereas those in the heavens underwent motion in circles or combinations of circles. It is true that on Earth most inanimate objects come to rest and remain so unless acted upon by a force, but Newton realized that the coming to rest was due to the force of friction acting in opposition to the motion. His keen insight allowed him to imagine a body moving at constant velocity in space and to realize that it would not come to rest, but would continue to move forever in a straight line <u>without</u> the aid of a force.

3.2 Newton's Second Law: Motion With a Force

Newton's second law states that when a force acts on a body, the body moves such that the force equals the mass times acceleration:

$$\vec{F} = m\vec{a} \qquad\qquad \text{(Eq. 3.2-1)}$$

where the arrow over the symbols F and a indicates that both these quantities are <u>vectors</u> and have both a <u>magnitude</u> and a <u>direction</u>. Force is measured in Newtons (N). Mass is a measure of the amount of matter contained in a body and is measured in kilograms (kg). It is a constant property of a body, independent of the body's environment. Acceleration may be due to a change in speed or a change in direction or both, and is measured in meters/second/second or meters/second^2 (m/s^2). Some examples are given in Table 3.2-1.

Table 3.2-1 Examples of Accelerated Motion

Speed	Direction	Example
Changing	Constant	Starting and stopping at a stoplight
Constant	Changing	Circular motion
Changing	Changing	Elliptical motion

In order to use Newton's second law to calculate the motion of a body, we need to put in what the force is: a constant or some function of time or the coordinates. Then we can rewrite Eq. 3.2-1 as:

$$\vec{a} = \vec{F}/m \qquad \text{(Eq. 3.2-2)}$$

If we know the direction of F and a, we may omit the vector signs and solve for the magnitude a.

Example Problem 3.2-1: A block of mass 2 kg is pushed with a force of 10 N. Find its acceleration.

$$a = F/m$$
$$a = 10 \text{ N} / 2 \text{ kg}$$
$$\boxed{a = 5m/s^2.}$$

In circular motion, the direction of the acceleration is always toward the center of the circle, i.e. it is a centripetal acceleration. Its magnitude may be easily derived as follows: In one complete orbit, the velocity vector changes direction by 2π radians:

change in velocity: $\Delta v = 2\pi v$

This change occurs in a time equal to the period of the orbit which is equal to the circumference divided by the speed:

elapsed time: $\Delta t = $ period $P = 2\pi r/v$

Thus, the magnitude of the acceleration is:

$$a = \Delta v/\Delta t = 2\pi v \times v/2\pi r$$
$$a = v^2/r \qquad \text{(Eq. 3.2-3)}$$

Newton's analysis of circular motion illustrates another major difference with previous ideas. In the Aristotelean view, celestial bodies undergo circular orbital motion without a force. In Newton's view, circular (or elliptical) orbital motion has an acceleration and requires a force to sustain it. That force is the attractive force of gravity.

3.3 Newton's Third Law

Newton's third law states that when two bodies A and B interact, the force $\vec{F}_{AB}$ exerted on body A by body B is equal in magnitude and opposite in direction to the force $\vec{F}_{BA}$ exerted on body B by body A.

$$\vec{F}_{AB} = -\vec{F}_{BA} \qquad \text{(Eq. 3.3-1)}$$

$\vec{F}_{AB}$ and $\vec{F}_{BA}$ are called an action–reaction pair and Table 3.3-1 gives some examples.

Table 3.3-1 Examples of Action–Reaction Pairs

Action	Reaction
Person pushing a cart	Cart pushing back on person
Person pressing down on scales	Scales pressing up on person
Earth exerting gravitational force on person	Person exerting gravitational force on Earth
Rearward thrust of rocket fuel	Forward motion of rocket

It may at first seem surprising that, in the case of a person and the Earth interacting gravitationally, the forces are equal, but such is indeed the case. If we combine the second and third law we get:

$$m_A a_A = m_B a_B \qquad \text{(Eq. 3.3-2)}$$

or equivalently:

$$a_B = m_A a_A / m_B \qquad \text{(Eq. 3.3-3)}$$

It is the acceleration of bodies, i.e. their change in velocity, which we observe and not the forces. According to Eq. 3.3-3, if body B has a much larger mass than body A ($m_B \gg m_A$), it will have a much smaller acceleration ($a_B \ll a_A$) which may be unobservable.

Example Problem 3.3-1: The acceleration of a person (or any other object) near the surface of the Earth is 9.8 m/s^2 due to the Earth's gravity. Calculate the resulting acceleration of the Earth due to the gravity of a 75 kg person.

Let A be a person and B be the Earth

$$a_B = m_A/m_B \times a_A$$
$$a_B = 75 \text{ kg} \times 9.8 \text{ m/s}^2 / 5.974 \times 10^{24} \text{ kg}$$
$$= 75 \times 9.8 / 5.794 \times 10 / 10^{24} \times \text{m/s}^2$$
$$a_B = 12.3 \times 10^{-23} \text{ m/s}^2$$

$$\boxed{a_B = 1.23 \times 10^{-22} \text{ m/s}^2}$$

An important consequence of Newton's third law is that bodies having large mass in a given system may be regarded as fixed because their accelerations are so small. Thus, the Sun may usually be regarded as fixed when considering the motions of the planets because it is at least 1000 times as massive as any of them.

3.4 Newton's Law of Gravity and Kepler's Third Law

Newton derived Kepler's laws of motion by assuming a universal law of gravitation and applying his laws of motion. He tried different forms of the law of gravitation, but the one that worked is the inverse square law, in which the gravitational force between two masses M and m separated by a distance r is attractive, i.e. directed toward the other mass, and has a magnitude given by:

$$F = G \times Mm/r^2 \qquad \text{(Eq. 3.4-1)}$$

where G is the universal gravitational constant.

Let us derive Kepler's third law for a planet of mass m orbiting the Sun of mass M at a distance r. We write down Newton's second law for the planet:

$$F = ma$$

and then substitute the gravitational force for F and the centripetal acceleration of circular motion for a:

$$G \times Mm/r^2 = mv^2/r$$

We may cancel out m and one power of r

$$G \times M/r = v^2$$

Now we use the relation between speed, circumference and period, $v = 2\pi r/P$, to obtain

$$G \times M/r = 4\pi^2r^2/P^2$$

and by rearranging the remaining terms we obtain

$$P^2 = (4\pi^2/GM) \times r^3 \qquad \text{(Eq. 3.4-1)}$$

which is Kepler's third law with $k = 4\pi^2/GM$.

Note that, if the law of gravity involved another power of r, e.g. r or 1/r, there would be a different relation between the period and the radius of the orbit. Newton <u>assumed</u> an inverse square law form for the law of gravitation and showed that Kepler's third law followed from it. It was not until after Newton, in the eighteenth century, that the inverse square law of gravitational attraction was directly verified in the laboratory by Cavendish.

Newton also showed by means of calculus that Kepler's third law holds for elliptical orbits as well, where the <u>semimajor axis</u> replaces the radius in Eq. 3.4-1.

3.5 Energy and Escape Velocity

Newton also introduced the concept of energy, which has two parts. The energy of motion, E_K, called kinetic energy, for a body of mass m having a speed v is given by:

$$E_K = \tfrac{1}{2} mv^2 \qquad \text{(Eq. 3.5-1)}$$

The energy of interaction, E_P, called potential energy, depends on the coordinates, and for a body of mass m located at a distance r from a body of mass M is given by:

$$E_P = - GMm/r \qquad \text{(Eq. 3.5-2)}$$

where the negative sign means that the mass m is in a potential well around mass M and positive energy must be added to break out of it. For a gravitational field, the total energy is conserved so that:

$$E_K + E_P = E = \text{constant.} \qquad \text{(Eq. 3.5-3)}$$

We may use the principle of <u>conservation of energy</u> to derive a formula for the escape velocity from a planet, i.e. the minimum velocity of a body which, when directed away from the planet, permits the body just to escape the planet's gravity.

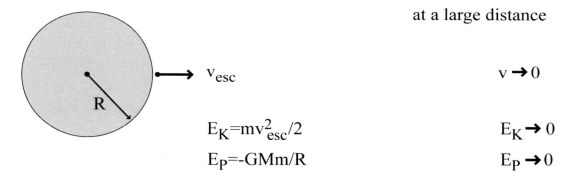

Fig. 3.5-1. Escape Velocity from a Planet.

The total energy is clearly zero at a large distance and must therefore be zero everywhere.

$$\tfrac{1}{2} mv_{esc}^2 - GMm/R = 0$$
$$\tfrac{1}{2} v_{esc}^2 = GM/R$$
$$v_{esc} = \sqrt{(2GM/R)} \qquad \text{(Eq. 3.5-4)}$$

Calculating the escape velocity of a planet is more than an exercise in Newtonian physics. It can provide some idea of what a planet's atmosphere might be like. Small planets or satellites having escape velocities of only a few km/s cannot retain their atmospheres for a very long time because a significant number of molecules are moving faster than the escape velocity and actually do escape. Large planets having escape velocities of tens of km/s hold onto all gases including hydrogen and helium and become gaseous giants. The Earth is in the middle, with an escape velocity of 11 km/s, and has lost the hydrogen and helium, but retained the nitrogen, oxygen and water vapor, which are essential for life.

3.6 Unity of Motion

Newton's laws of physics beautifully unified types of motion which had previously been thought to be quite different. Imagine a large cannon that can shoot a projectile at a variable tangential speed v_t parallel to the Earth's surface. Four possible paths are shown in Fig. 3.6-1.

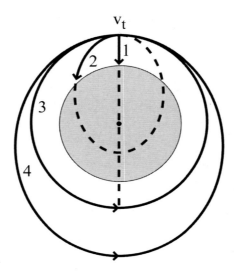

Fig. 3.6-1. Falling and Orbital Motion.

If v_t is zero, the projectile drops straight down (Path 1). This motion is <u>falling</u>. If v_t is non-zero, but less than the circular orbital speed $v_o = \sqrt{(GM/R)}$ (8 km/s for the Earth), the projectile moves in a curved path which eventually strikes the Earth (Path 2). This motion is called <u>projectile motion</u>. If v_t is equal to v_o, the projectile undergoes <u>circular orbital motion</u> (Path 3). If v_t is greater than v_o, the projectile undergoes <u>elliptical orbital motion</u> with its point of origin at perigee (Path 4).

In Newton's view, the four different trajectories depicted in Fig. 3.6-1 all belong to the class of allowed orbits in a gravitational field and differ only in their <u>tangential speeds</u>. In fact, if the Earth (assumed here to be a sphere) were replaced by a point at its center which had the same mass, its gravitational effect on the projectile would be the same. In this case, without the solid Earth in the way, the projectile following Path 2 would complete an elliptical orbit (dashed line) with its point of origin at apogee. The projectile following Path 1 would complete a straight-line orbit directly through the center. Bodies near the Earth and in the heavens obey the same laws of physics and follow the same set of paths.

Questions

3-1. What is the key difference between Newton's idea of motion and the earlier idea espoused by Aristotle?

3-2. How can we calculate the motion of a body by using the equation $F = ma$?

3-3. When are the accelerations of two interacting bodies, A and B, equal?

3-4. Under what conditions may a body be regarded as fixed?

3-5. How did Newton deduce that the force of gravity was an inverse-square law?

3-6. What basic principle of physics is used in deriving the formula for escape velocity?

3-7. What does the escape velocity of a planet tell us?

3-8. In Newton's view, what quantity distinguishes different types of motion such as falling, circular motion or elliptical motion?

Problems

3-1. Calculate the acceleration of a person standing near the edge of a merry-go-round at a distance of 3 meters from the center and traveling at a speed of 5 meters per second.

3-2. Two satellites A and B have the same density. B has twice the radius of A. What is the escape velocity of B relative to that of A?

3-3. In some regions of the disk of the Milky Way, the centrally-directed gravitational force falls off inversely as the distance from the center, i.e. $F = k/r$. How does the orbital speed v depend on r?

3-4. A projectile is launched horizontally from a high tower on Earth with a speed of 10 km/s. Which path in Fig. 3.6-1 will it follow?

Chapter 4: The Earth

The Earth is the planet on which we live and about which we have the most detailed knowledge. It serves as a basis for our investigation about the other planets, satellites, and smaller bodies in the solar system.

4.1 Surface Gravity, Rotation, and Shape

When a body of mass m is on the surface of a planet of mass M and radius R, the gravitational force F_W, which it feels from the planet and which is its weight, is given by Newton's universal law of gravitation as:

$$F_W = m\,(GM/R^2) \qquad\qquad \text{(Eq. 4.1-1)}$$

The quantity in parentheses is the surface gravitational acceleration or surface gravity and is denoted by g:

$$g = GM/R^2 \qquad\qquad \text{(Eq. 4.1-2)}$$

The weight of a body is equal to its mass times the surface gravity:

$$F_W = mg \qquad\qquad \text{(Eq. 4.1-3)}$$

Example Problem 4.1-1: Calculate the surface gravity g_e at the equator of the Earth, using the equatorial radius R_e of 6.378×10^6 m.

$g_e = GM/\,R_e^{\,2}$
$g_e = 6.67259 \times 10^{-11}\ \text{m}^3/\text{kg/s}^2 \times 5.974 \times 10^{24}\ \text{kg} / (6.378 \times 10^6\ \text{m})^2$
$g_e = 6.67259 \times 5.974 / (6.378)^2 \times 10^{-11+24-12}\ \text{m/s}^2$

$\boxed{g_e = 9.799\ \text{m/s}^2.}$

Because the Earth is rotating fairly rapidly, material near the equator feels an outward centrifugal force and creates an equatorial bulge so that the equatorial radius is 21 km more than the polar radius.

Example Problem 4.1-2: Calculate the surface gravity g_p at the poles of the Earth, using the polar radius R_p of 6.357×10^6 m. Calculate the percent difference of g_e from above with respect to g_p.

Proceeding in a similar manner to above we obtain:

$g_p = 9.864\ \text{m/s}^2$
% difference $= (g_e - g_p)/g_p \times 100\%$
$= (9.799 - 9.864) / 9.864 \times 100\%$

$\boxed{\text{\% difference} = -\,0.66\%.}$

At the equator, the same centrifugal force which produces the equatorial bulge acts oppositely to the gravitational force and effectively reduces the surface gravity. The centripetal acceleration is given by

$$a_c = \omega^2 R_e \qquad \text{(Eq. 4.1-4)}$$

where ω is the angular velocity of rotation of the earth (= 2π radians / sidereal period) and the sidereal period is 86160 seconds. We then subtract a_c from g_e calculated above to obtain the effective surface gravity at the equator g_e'.

$$g_e' = g_e - a_c \qquad \text{(Eq. 4.1-5)}$$

<u>Example Problem 4.1-3:</u> Calculate for the equator the centripetal acceleration a_c, the effective surface gravity g_e', and the percent difference from g_p.

$$a_c = (2\pi)^2 / (8.616 \times 10^4 \, \text{s})^2 \times 6.378 \times 10^6 \, \text{m}$$

$$\boxed{a_c = 3.4 \times 10^{-2} \, \text{m/s}^2}$$

$$g_e' = g_e - a_c$$
$$g_e' = 9.799 \, \text{m/s}^2 - .034 \, \text{m/s}^2$$

$$\boxed{g_e' = 9.765 \, \text{m/s}^2}$$

% difference = $(g_e' - g_p)/g_p \times 100\%$
% difference = $(9.765 - 9.864) / 9.864 \times 100\%$

$$\boxed{\% \text{ difference} = -1.00\%.}$$

Thus, you would weigh almost exactly 1% less at the equator than at the poles.

This calculation gives the answer to the question posed by the ancient and medieval astronomers about the tendency of things to go flying off the Earth if the Earth spun on its axis. This tendency exists but the <u>centrifugal</u> acceleration is at most only about 1% of the <u>centripetal</u> acceleration. Gravity is so strong that Earth holds on to its objects even though it is rotating fairly rapidly.

4.2 Plate Tectonics

The idea of continental drift was first championed by the German meteorologist and polar explorer Alfred Wegener in 1915 in his book <u>The Origins of the Continents and the Oceans</u>. His main evidence was the existence of similar fossils widely distributed across the globe, whose distribution could be explained by the theory that they originated on a supercontinent which subsequently broke apart and whose fragments drifted to widely separated positions. However, the theory, although intriguing, was rejected by most geologists for decades because they could not imagine a force powerful enough to overcome the enormous friction as the continents "plowed through" the ocean basins.

In the 1960s, following years of accumulating evidence that continental drift really did occur, geologists finally realized that the continents and ocean basins could move together on plates which extend to a depth of 100 km, where the rock is partially melted and the lower friction permits much easier sliding. Thus, the idea of plate tectonics was born. We now know that the Earth's surface consists of about a dozen large plates together with many smaller ones, which are in constant motion at rates varying between 1 and 10 centimeters per year. Table 4.2-1 lists the three main types of plate boundaries.

Table 4.2-1 The Three Main Types of Plate Tectonics

Boundary Type	Relative Motion	Example
Rift zone	← →	Mid-Atlantic ridge
Subduction zone	→ ←	Off the coast of Japan
Fault zone	→ ←	San Andreas fault

Most seismic (earthquake) activity and volcanic activity occurs at or near plate boundaries, but there are some significant active areas in the interior of plates. These are mainly of two types and are listed in Table 4.2-2.

Table 4.2-2 The Two Main Types of Interior Active Areas

Area Type	Cause	Activity Type	Example
Hot spot	Magma plume in the mantle	Volcanoes	Hawaiian islands
Weak spot	Imperfect "weld" in the plate	Earthquakes	Near New Madrid, Missouri

In the Hawaiian Islands, the magma plume lies under the Pacific plate which is moving northwest at several centimeters per year. The plume burns through the crust and vents out in a volcano for a while and then, when the plate drift has carried the volcano too far away, it burns through at another spot and vents there, etc. In this manner, a chain of volcanic islands is created.

The weak spot near New Madrid, Missouri produced what is thought to have been the most powerful earthquake in the continental United States in historical times. Damage was minor because of the low population at the time, but a similar event today would cause widespread destruction from St. Louis, Missouri to Memphis, Tennessee.

No place on the surface of the Earth is entirely earthquake free because the plates are moving everywhere. An ideal location is not one which has (or rather appears to have!) no earthquakes at all. What this means is that the stresses are building up in the rocks and at some time a catastrophic earthquake will occur. It is better to live with minor earthquakes from time to time which serve to relieve the stress more gradually and prevent the "big one" from occurring.

Although some progress has been made in predicting small and moderate earthquakes, the ones we would most like to predict, i.e. the catastrophic ones, by their very nature, fail to give easily detectible warning of their onset. However, there is much we can do in the area of damage mitigation. We know enough about how to construct buildings, bridges, and other structures to withstand most earthquakes and minimize damage and loss of life. The challenge is to implement this knowledge.

Earthquakes generate seismic waves, whose propagation through the Earth provides us with our knowledge of the Earth's interior. They are of two basic types shown in Table 4.2-3.

Table 4.2-3 The Two Types of Seismic Waves

Wave Type	Vibration (← →) and Propagation	Travel through Liquid?
S waves	$\updownarrow \Rightarrow$	No
P waves	$\rightarrow \Rightarrow$	Yes

S waves cannot travel through a liquid because when a liquid is displaced side-to-side, there is no <u>restoring force</u> to bring it back as there is a in a solid. The Earth's liquid outer core creates a shadow zone for S waves near the point on the Earth directly opposite the origin of the waves, and its size reveals the size of the liquid core.

4.3 Tsunamis

The compound word 'tsunami' in Japanese means 'harbor wave'. Tsunamis are large and often catastrophically destructive ocean waves created by earthquakes or volcanoes. In either case, when a large amount of undersea rock is suddenly displaced, a large wave having a wavelength of hundreds of kilometers is generated. In mid-ocean, the amplitude is so small (< 1 m) and the slope of the wave so gentle that it is not noticeable from a ship, but in bays and harbors the shallower depth and narrowing width focus the wave and build it to great heights. The energy of all this moving water is enormous.

<u>Example Problem 4.3-1</u>: Calculate the kinetic energy of a tsunami having a height h of 10 m, a width w of 2 km, a length l of 5 km, and moving at a speed v of 10 m/s².

$$E_k = \frac{1}{2} mv^2$$
$$E_k = \frac{1}{2} \rho hwlv^2$$
$$E_k = \frac{1}{2} \times 10^3 \text{ kg/m}^3 \times 10\text{m} \times 2 \times 10^3 \text{ m} \times 5 \times 10^3 \text{ m} \times (10 \text{ m/s})^2$$
$$E_k = \frac{1}{2} \times 2 \times 5 \times 10^{3+1+3+3+4} \text{ kg m/s}^2$$
$$\boxed{E_k = 5 \times 10^{14} \text{ J.}}$$

The best defense against tsunamis is to have in place a good warning system to alert people to leave the beaches and the low-lying areas. We have satellite systems which can detect the presence of the mid-ocean waves generated by earthquakes and volcanoes within minutes. The challenge lies in getting the information to the beaches and coastal areas quickly. Japan has lived with tsunamis for years and has a good system. There was no such system in the Indian Ocean on December 26, 2005, and the tsunami triggered by an undersea earthquake off the coast of Indonesia devastated the coasts of southern Asia and took over a quarter million lives. The world now realizes the importance of having a good warning system for every ocean basin.

4.4 The Coriolis Force

The Coriolis force is an apparent force felt in the rotating system of the Earth by any object moving with respect to the Earth. As seen looking in the direction of the object's motion, the deflection is to the right in the northern hemisphere and to the left in the southern hemisphere. The magnitude of the Coriolis acceleration for an object of mass m traveling at speed v is:

$$a_{co} = 2\omega v \sin\theta \qquad \text{(Eq. 4.4-1)}$$

where ω is the angular velocity of the Earth (= 2π radians / period, period = 86160s) and θ is the latitude. The Coriolis acceleration vanishes at the equator. The corresponding force is simply the mass times acceleration:

$$F_{co} = ma_{co} \qquad \text{(Eq. 4.4-2)}$$

Example Problem 4.4-1: Calculate the Coriolis acceleration on a car traveling at a speed of 100 km/hr at a latitude of 45°:

$$v = 100 \text{ km/hr} = 10^5 \text{ m/ } 3600s = 27.8 \text{ m/s}$$
$$a_{co} = 2 \times 2\pi/(8.616 \times 10^4 \text{ s}) \times 27.8 \text{ m/s} \times \sin45°$$
$$a_{co} = 2 \times 2\pi \times 27.8 \times .707 / 8.616 \times 10^{-4} \times \text{ m/s}^2$$
$$\boxed{a_{co} = 2.87 \times 10^{-3} \text{ m/s}^2}$$

Note that a_{co} is very small compared to g.

When low pressure areas form, the surrounding air flows in to fill the partial vacuum. However, it feels the Coriolis force and so does not flow straight in, but in a spiral path. In the northern hemisphere the rightward deflection yields a counterclockwise spiral, whereas in the southern hemisphere the leftward deflection leads to a clockwise spiral. Near the equator, where the Coriolis force is weak, there is a dead zone for rotating weather systems. Hurricanes and tropical storms do not form within 10° of the equator.

In high pressure systems the surface airflow is away from the center and the direction of flow is just reversed from low pressure systems, i.e. clockwise in northern hemisphere and counterclockwise in southern hemisphere.

Questions

4-1. What two factors contribute to making the surface gravity of the Earth smaller at the equator than at the poles?

4-2. Why do not objects at the equator fly off the Earth due to centrifugal force?

4-3. Why was Alfred Wegener's theory of continental drift rejected by most geologists for decades?

4-4. How are the plates which comprise the Earth's surface able to move with respect to one another?

4-5. What are the three main types of plate boundaries? Give one example of each, other than those in Table 4.2-1.

4-6 How is a chain of volcanic islands like the Hawaiian islands created?

4-7. Why is it not safe to live in an area which (for a time) has no earthquakes at all?

4-8. What can we do to minimize damage and loss of life from earthquakes?

4-9. If you were on a beach and the water suddenly went out much further than the normal low-tide level, what should you do?

4-10. In what directions are the airflows around low and high pressure areas in the northern and southern hemispheres?

Problems

4-1. Calculate the surface gravity of Mars, using a radius of 3394 km and a mass of 6.4×10^{23} kg. By how much is this reduced at the equator due to Mars' rotation? Its rotation period is 24 hours and 37 minutes.

Chapter 5: The Moon

The Moon is our nearest neighbor and the only celestial body on which mankind has set foot. Yet, despite its being so close and familiar, its low density poses a major problem in trying to explain its formation. The several cyclical changes in its orbit exemplify the effects of a perturbing third body, namely the Sun, on a two-body elliptical orbit. The changes over time of the rotation and revolution of the Earth and the Moon typify the evolution of a planet–satellite system.

5.1 The Formation of the Moon

The principal observational fact which any theory of the formation of the Moon must explain is the large difference between its density (3300 kg/m^3) and that of the Earth (5500 kg/m^3). The significantly lower density of the Moon means that the Moon cannot have much iron and nickel in its core and cannot be just a scaled down version of the Earth. Over the years, four basic theories of lunar formation have been proposed. They are listed in Table 5.1-1 together with their major problems (if any).

Table 5.1-1. Theories of Lunar Formation and Their Major Problems.

Theory	Major Problem
Capture theory	Needs a third body; very unlikely
Twin formation theory	Cannot explain the density difference
Fission theory	Requires too much angular momentum for the Earth–Moon system
Giant impact model	None

The capture theory states that the Moon formed at a different location and developed a different density but was then ejected and captured into orbit by the Earth. The problem is that a third body would have to be in just the right place at just the right time, and the probability of such an occurrence is very small. The capture process is important in astronomy sometimes, e.g. in the case of comet nuclei coming from the Oort cloud captured by one of the Jovian planets into the solar system, but does not seem appropriate for the Earth–Moon system.

In the twin formation theory, the Earth and Moon formed together in the Earth's orbital zone around the Sun. However, their resulting compositions should have been very similar, so this theory cannot explain the density difference.

In the fission theory, the young Earth rotated so fast that it spun off part of its outer material (having a lower density) to form the Moon. The major problem is that the total angular momentum of the Earth–Moon system must remain constant, and if the Earth had been spinning fast enough in the past for fission to occur, it would still be spinning about four times faster than it is. The fission process is important in astronomy sometimes, e.g. in the case of the formation of close binary stars but does not seem appropriate for the Earth–Moon system.

The giant impact model starts out like the twin formation theory and then the smaller body, which is about the size of Mars, collides with Earth and knocks off some of its outer material (having lower density), some of which goes into Earth orbit and reaccretes to form the Moon. This model has no major problems. The ultimate collision of two bodies in close orbit would be a near certainity and the reaccretion process has been simulated by computer modeling. The giant impact model remains the best theory of lunar formation proposed to date.

5.2 The Moon's Orbit

A Two-Body System

If the effects of the Sun and the other planets are, for the moment, neglected, the Earth–Moon system may be regarded as an orbiting two-body system. As the Moon orbits the Earth in response to the Earth's gravitational force, the Earth does the same in response to the Moon's gravitational force, albeit in a smaller orbit. According to Newton's third law, the gravitational forces on the two bodies are equal in magnitude but opposite in direction, so the relative accelerations, velocities and positions of the Earth and the Moon are in inverse proportions to their respective masses, so that Earth's values are $M_M/M_E = 1/81.3$ times those of the Moon.

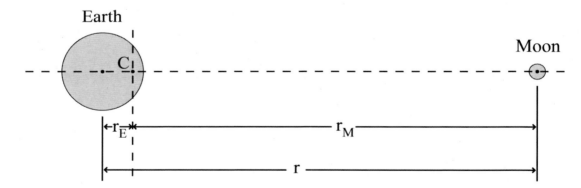

Fig. 5.2-1. The Earth–Moon System and its Center of Mass.

If r is the variable distance between the centers of the Earth and the Moon, the center of mass (also called the barycenter) is located at point C such that the center of the Earth is at a distance

$$r_E = M_M/(M_E + M_M) \times r \qquad \text{(Eq. 5.2-1)}$$

and the center of the Moon is at a distance

$$r_M = M_E/(M_E + M_M) \times r \qquad \text{(Eq. 5.2-2)}$$

Example Problem 5.2-2: Calculate r_E for the Earth-Moon system using the values $M_E = 5.974 \times 10^{24}$ kg, $M_M = 7.349 \times 10^{22}$ kg and $r = 3.843 \times 10^5$ km.

$$r_E = M_M/(M_E + M_M) \times r$$
$$M_M = .07349 \times 10^{24} \text{ kg}$$
$$r_E = .07349/(5.974 + .07349) \times 3.843 \times 10^5 \text{ km}$$

$$\boxed{r_E = 4.67 \times 10^3 \text{ km.}}$$

r_E is smaller than the radius of the Earth so that the center of mass is inside the Earth. There are thee ellipses relevant to the orbiting Earth–Moon system. They are identical in shape but have a different semimajor axis depending on viewpoint. They are listed in Table 5.2-1.

Table 5.2-1. The Three Ellipses of the Earth–Moon System.

Viewpoint	Orbiting Body	Semimajor Axis
Center of Earth	Moon	a
Center of Moon	Earth	a
Center of Mass	Earth	$[M_M/(M_E + M_M)]$ a
Center of Mass	Moon	$[M_E/(M_E + M_M)]$ a

Note that as viewed from the center of mass the Earth's orbit is just like the Moon's but it is smaller by a factor of M_M/M_E (1/81.3).

Kepler's laws are still valid for the orbiting two-body system, but in the third law we must replace the single mass M in Eq. 3.4-1 by the sum of the masses.

$$P^2 = 4\pi^2/G(M_E + M_M) \times a^3 \qquad \text{(Eq. 5.2-3)}$$

We use this equation, with two stellar masses M_A and M_B, to determine the total mass of binary star. In this case, the masses and the size of the orbits are comparable (see Section 12.6).

Synchronous Rotation
The Moon orbits the Earth in synchronous rotation, which means that its rotation and revolution periods are equal and have the value of 27.322 days, which is called the sidereal month. This equality of rotation and revolution periods has not always been the case. Soon after its formation, the Moon was spinning more rapidly, but the strong tidal force of the Earth has slowed it down until it keeps the face toward the Earth. This synchronism holds exactly when averaged over one orbit. However, because the angular velocity of revolution varies along the elliptical orbit, rotation alternately gets ahead and then behind revolution

during one orbit. Therefore, the part of the Moon which we see changes back and forth by several degrees each way, and overall we can see about 60% of the lunar surface from the Earth.

From any location on the Moon, the Earth would move back and forth by several degrees during each orbit but, on average, remain fixed in the sky.

Phase Period versus Revolution Period

The period of the Moon which is most familiar and important to us is not the revolution period of 27.322 days with respect to the stars but the phase period, i.e. the time from one full moon to the next or one new moon to the next. Because the Earth is orbiting the Sun as the Moon is orbiting the Earth, the Moon has to move more in its orbit to line up with the Sun again, and the phase period is approximately two days longer at 29.53059 days, which is called the <u>synodic month</u>. This length of month is the value that is used in lunar calendars and is slightly shorter than the average value of 30.4375 days in the Gregorian calendar.

A striking sight for future lunar colonists would be an eclipse which is a lunar eclipse for terrestrial observers but a <u>solar eclipse</u> for lunar observers. While the Earth remained nearly stationary in the sky, the Sun would proceed across the sky at its slow pace of 12.19° per day. Its one-half degree disk would slip behind the much larger two-degree disk of the Earth, creating a bright ring of light around the edge of the Earth and bathing the moonscape in a dull coppery glow of light for one to two hours. Lunar colonists anywhere on the side facing the Earth would typically be able to see this sight every one to two years.

Periodic Variations in the Moon's Orbit: Apogee, Perigee and Eccentricity

If the Earth–Moon system were alone in the universe, the two bodies would continue to move indefinitely in elliptical orbits whose dimensions, shapes and orientations remained constant. However, the Earth–Moon system is part of the solar system and is subject to perturbing forces from the Sun and other planets, the largest being from the Sun. As the Moon orbits the Earth and the Earth orbits the Sun, the Sun pulls on the Moon, shifting the position of apogee and perigee back and forth and alternately increasing and decreasing the eccentricity of the orbit. Those changes are approximately cyclic and one of these eccentricity-phase cycles is shown in Fig. 5.2-2.

For simplicity, we assume that when the Earth–Moon system is at Position 0, the Moon is at perigee and also at new moon, so that its phase and orbital positions coincide. In this position, the eccentricity has a maximum value, averaging near .065, and apogee and perigee are close to being on opposite sides of the orbit, i.e. the apogee-to-perigee angle is close to 180°. The Sun's pull causes the position of perigee along the orbit (longitude) to advance more rapidly than that of apogee so that the apogee-to-perigee angle becomes greater than 180°, reaching a maximum of about 210° near Position 1 and decreasing to 180° near Position 2. At Position 2, the eccentricity reaches a minimum value, averaging near .045. From Position 2 to near Position 3, the longitude of apogee advances more rapidly than that of perigee so that the apogee-to-perigee angle becomes less than 180°,

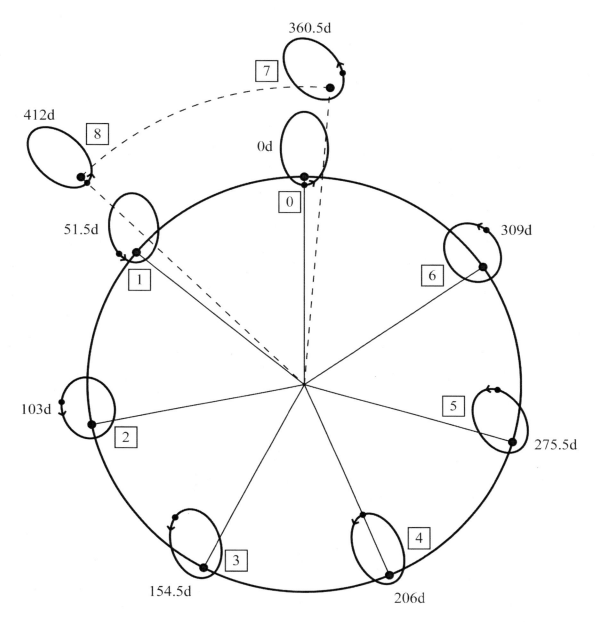

Fig. 5.2-2. The Axis-Phase-Eccentricity Cycle of the Moon's Orbit.

reaching a minimum of about 150° near Position 3 and then increasing to 180° near Position 4. At Position 4, the eccentricity again reaches a maximum value, averaging near .065. At this point, the eccentricity has completed the full cycle but the Moon is at apogee instead of perigee, having completed 7 perigee-to-perigee orbits plus a perigee-to-apogee half orbit in 7 synodic months. Although apogee and perigee have undergone both advances and retreats along the orbit, there has been a net advance for both so that the major axis of the orbit is rotated by 23.0° compared to its starting position, and the Earth has traveled more than halfway around in its orbit. This process from Position 0 to Position 4 constitutes the lunar eccentricity cycle of approximately 206 days.

The variations in apogee, perigee and eccentricity are repeated from Position 4 through Positions 5, 6 and 7 to Position 8. Now, the Moon is again at perigee and the Sun, Earth and Moon have the same alignment as originally but the Earth–Moon system is 46.0° further around in the Earth orbit. The entire process from Position 0 to Position 8, comprising 15 perigee-to-perigee orbits in 14 synodic months constitutes the lunar phase-eccentricity cycle of approximately 412 days. The time required for a perigee-to-perigee (or equivalently apogee-to-apogee) orbit, i.e. 27.55457 days which is slightly longer than the sidereal month because of the rotation of the major axis is called the <u>anomalistic month</u>. The time required for the major axis to make a complete rotation of 360° is 8.82 years.

The variation in eccentricity for the period 2004–2006 is shown in Fig. 5.2-3 for the three-year period 2004–2006. The curve is basically a sinusoidal curve whose only departure from an ideal sinusoid is that the time required to go from the average value through a minimum and back is less by a factor of 1.4 than the corresponding time for a maximum. This difference arises because the net effects of the perturbing forces on the eccentricity are stronger near the minima than near the maxima. Also note that at any given time, the eccentricity of the Moon's orbit is unlikely to have a value near its average value of .055.

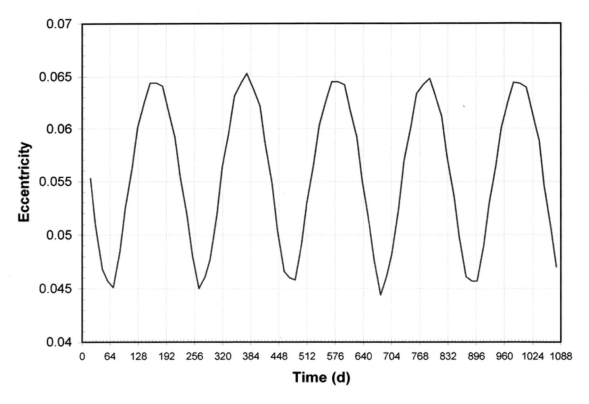

Fig. 5.2-3. Eccentricity of the Moon's Orbit: 2004–2006.

The procedure for determining the characteristics of the Moon's orbit from changes in its apparent diameter is illustrated in Laboratory Exercise LE1 for two different periods, one near an eccentricity minimum and the other one near an eccentricity maximum.

Periodic Variations in the Moon's Orbit: The Plane of the Orbit

The cyclic variations in the Moon's orbit discussed above would occur in a nearly identical manner if the lunar orbit were in the plane of the ecliptic. In fact, the plane of the orbit is inclined by 5.2° to the ecliptic plane. As a result, the Sun exerts a torque on the orbiting Earth-Moon system which causes the orbital plane to precess in much the same manner that gravity exerts a torque on a spinning gyroscope and makes it precess when it is placed on a stand. This period of precession is 18.6 years and is different from the major-axis precession period discussed above and the direction is opposite. This cycle gives rise to the saros cycle of the solar eclipses, in which an eclipse very similar to a given one occurs 18 years and 11 days later, one third of the way around the Earth.

Summary of the Cycles of the Moon's Orbit

We have discussed the five principle cycles of the Moon's orbit; three months and two longer periods. They are summarized in Table 5.2-1.

Table 5.2-1. The Principle Cycles of the Moon's Orbit.

Cycle	Period	Meaning
Sidereal Month	27.322 days	Revolution of Moon with respect to stars
Anomalistic Month	27.55457 days	Revolution of Moon: apogee to apogee or perigee to perigee
Synodic Month	29.5306 days	Revolution of Moon: new to new or full to full
Major-Axis Precession Cycle	8.82 years	Rotation of major axis with respect to stars
Orbital Precession Cycle	18.6 years	Rotation of plane of orbit with respect to stars

5.3 The Tides

Tidal bulges are produced in/on a body when there is a significant difference in the gravitational force acting at different points in/on it from a nearby gravitating body. The situation for the Earth–Moon or Earth–Sun system is shown in Fig. 5.3-1.

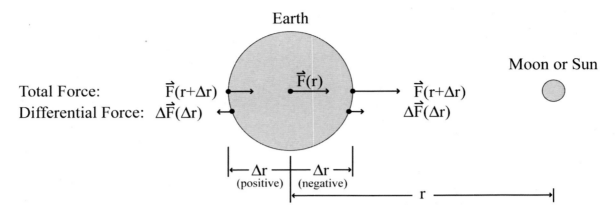

Fig. 5.3-1. Tidal Forces on the Earth from the Moon.

If the distance between the center of the Earth and the center of the Moon or Sun is r, then the gravitational force on a mass m at the center of the Earth is:

$$F(r) = GM_M m/r^2 \qquad\qquad \text{(Eq. 5.3-1)}$$

Consider the gravitational force on the same mass m at a distance $r + \Delta r$, where Δr may be a positive or negative quality. It is

$$F(r + \Delta r) = GM_M m/(r + \Delta r)^2 \qquad\qquad \text{(Eq. 5.3-2)}$$

Δr is small compared to r so the $\Delta r^2/r^2$ term may be neglected.

$$F(r + \Delta r) = (GM_M m/r^2)(1 - 2\Delta r/r) \qquad\qquad \text{(Eq. 5.3-3)}$$

What counts in producing tidal effects is not the total force but the differential force ΔF relative to the force at the center $F(r)$. That force constitutes the force necessary to drive the Earth as a single mass in its elliptical orbit. Additional differential forces cause parts of the Earth to move differentially with respect to each other. The differential force is

$$\Delta F = F(r + \Delta r) - F(r) \qquad\qquad \text{(Eq. 5.3-4)}$$

which from equations 5.3-1 and 5.3-3 is

$$\Delta F = -2GM_M m\Delta r/r^3 \qquad\qquad \text{(Eq. 5.3-5)}$$

The tidal force falls off as the inverse cube of the distance from the tide-producing body and is directed toward the tide-producing body on the near side and away from the tide-producing body on the far side. These tidal forces create two tidal bulges, which produce the twice daily ocean tides as the Earth rotates.

It is evident from Eq. 5.3-5 that the relative strengths of the lunar and solar tides are proportional to the mass of the tide-producing body and inversely proportional to the cube of its distance from the Earth.

Example Problem 5.3-1: Calculate the relative strengths of the lunar and solar tides.

$$\Delta F_M / \Delta F_S = (M_M / M_S)\,(r_S{}^3 / r_M{}^3)$$
$$\Delta F_M / \Delta F_S = (7.349 \times 10^{22} kg\,/\,1.989 \times 10^{30} kg)\,(1.496 \times 10^8\ km / 3.843 \times 10^5\ km)^3$$
$$\Delta F_M / \Delta F_S = .218 \times 10^{-8} \times 10^9$$

$$\boxed{\Delta F_M / \Delta F_S = 2.18}$$

At spring tides, the lunar and solar tides add, whereas at neap tides they subtract. Theoretically, this ratio should be $(2.18 + 1)/(2.18 - 1) = 2.7$, although in practice it is often closer to 2.

5.4 Evolution of the Earth–Moon System

The full action–reaction tidal interaction of the Earth-Moon system is shown in Fig. 5.4-1.

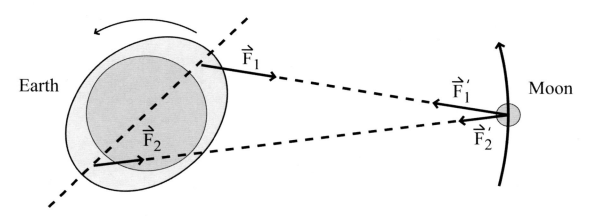

Fig. 5.4-1. Action–Reaction of the Earth-Moon Tidal Interaction.

Because of frictional forces on the tidal waters, the Earth's rotation carries the double tidal bulge forward of the Earth–Moon line. The Moon's gravity exerts a force $\vec{F}_1$ on the near-side bulge which tends to slow the Earth's rotation and a force $\vec{F}_2$ on the far-side bulge which tends to speed it up. However, the near-side bulge is slightly nearer to the Moon, so $\vec{F}_1$ is slightly larger and there is a net <u>slowing</u> of the Earth's rotation. By Newton's third law, the reaction forces $\vec{F}_1{}'$ and $\vec{F}_2{}'$ are equal in magnitude, but opposite in direction to $\vec{F}_1$ and $\vec{F}_2$, respectively. $\vec{F}_1{}'$ is slightly larger than $\vec{F}_2{}'$ so there is a net <u>forward force</u> in the direction of the Moon's motion which boosts it into a higher orbit, reducing its speed but increasing its potential energy and angular momentum.

The ultimate result of this interaction can be easily determined by consideration of the total angular momentum of the system, which consists of orbital and rotational parts for both the Earth and the Moon.

For simplicity, we assume that the mutual orbits of the Earth and the Moon are circles of radii r_E and r_M, respectively. Orbital angular momentum L is equal to mass time radius squared times orbital angular velocity so that the total for the system is

$$L = L_E + L_M = M_E r_E^2 \omega + M_M r_M^2 \omega \qquad \text{(Eq. 5.4-1)}$$

where ω = angular velocity of revolution = 2π/month. Since r_E is smaller than r_M by the mass ratio M_E/M_M (= 81.3), the orbital angular momentum of the Earth may be neglected. Rotational (spin) angular momentum S is equal to the moment of inertia (= $2MR^2/5$ for a sphere) times rotational angular velocity so that the total for the system is

$$S = S_E + S_M = 2M_E R_E^2 \Omega/5 + 2M_M R_M^2 \omega/5 \qquad \text{(Eq. 5.4-2)}$$

where Ω = angular velocity of Earth's rotation = 2π/day. In this case, the Moon's quantities are all smaller than the Earth's and its rotational angular momentum may be neglected. Therefore, to a very good approximation, the total angular momentum J of the Earth–Moon system is

$$J = S_E + L_M = 2/5 M_E r_E^2 \Omega + M_M r_M^2 \omega \qquad \text{(Eq. 5.4-3)}$$

Example Problem 5.4-1: Calculate the values of the Moon's orbital angular momentum and the Earth's rotational angular momentum in Eq. 5.4-3.

$$S_E = 2M_E R_E^2 \, \Omega/5$$
$$S_E = 2/5 \times 5.97 \times 10^{24} \text{ kg} \times (6.38)^2 \times 10^{12} \text{ m}^2 \times 7.25 \times 10^{-5} \text{ s}^{-1}$$

$$\boxed{S_E = 7.07 \times 10^{33} \text{ kgm}^2/\text{s}^2}$$

$$L_M = M_M r_M^2 \omega$$
$$L_M = 7.35 \times 10^{22} \text{ kg} \times (3.84)^2 \times 10^{16} \text{ m}^2 \times 2.66 \times 10^{-6} \text{ s}^{-1}$$

$$\boxed{L_M = 28.8 \times 10^{33} \text{ kgm}^2/\text{s}^2}$$

$$\boxed{L_M = 4S_E.}$$

It is one of the fundamental laws of physics that total angular momentum must remain constant. As a result of the tidal interaction, the Earth loses rotational angular momentum and the Moon gains orbital angular momentum by the same amount. The Moon can gain angular momentum from the Earth only until the Earth slows down enough so that the Earth–Moon system would have double synchronous rotation. The system then would be stable and there would be no more changes. At that point, the Moon would have 5/4 of the angular momentum it has now.

<u>Example Problem 5.4-2:</u> What would be the radius of the Moon's orbit and its period if the Earth–Moon system evolved to the point of double synchronous rotation.

Let the current orbit have a radius r_1, and angular velocity ω_1, and the doubly synchronous orbit have a radius r_2 and angular velocity ω_2

$$L_2/L_1 = M_M r_2^2 \omega_2 \,/\, M_M r_1^2 \omega_1 = r_2^2 \omega_2 / r_1^2 \omega_1$$

Since the gravitating body, namely the Earth, remains the same, the two orbits are related by Kepler's third law.

$$P_2/P_1 = \omega_1/\,\omega_2 = r_2^{3/2}/r_1^{3/2}$$
$$L_2/L_1 = r_2^{1/2}/r_1^{1/2}$$
$$r_2/r_1 = (L_2/L_1)^2 = (5/4)^2 = 25/16$$

$$\boxed{r_2 = 25/16 \; r_1 = 1.5625 \times r_1 = 6.00 \times 10^8 \text{ m}}$$
$$\boxed{P_2 = (25/16)^{3/2} \; P_1 = 1.953 \times P_1 = 53.3\text{d}}$$

Problem 5.4-2 is to estimate the time required for this evolution to occur.

In fact, the Sun's evolution into a red giant will not permit the current evolution of the Earth–Moon system to reach the doubly synchronous state. However, in principle, double synchrony is the ultimate fate of a two-body system with tidal interactions. There is at least one two-body system in our solar system, namely the Pluto–Charon system, where it actually has occurred.

Questions

5-1.   What are the four theories of the formation of the Moon and the problems (if any) with each of them?

5-2.   What is the ratio of the accelerations, velocities and positions (with respect to the center of mass) for the Earth–Moon system?

5-3.   What change must be made in Kepler's third law for an orbiting two-body system?

5-4.   Why does the part of the Moon visible from the Earth vary slightly over the course of a month?

5-5.   Why is the phase period of the Moon different from the sidereal period?

5-6.   What causes the periodic variations in apogee, perigee and eccentricity of the Moon's orbit?

5-7.   What are the most likely values of the eccentricity of the Moon's orbit?

5-8.   How many principle cycles of the Moon's orbit are there? List them and give a brief definition for each.

5-9.   How does the strength of the tidal force depend on the distance of the tide-producing body?

5-10. Which type of motion for which body contains most of the angular momentum of the Earth–Moon system?

Problems

5-1.   When the eccentricity of the Moon's orbit is a maximum (e = .065), what is the ratio of apogee to perigee distances? (See Appendix A2). What is the ratio of apogee-to-perigee tidal forces? (Tidal force is proportional to the inverse cube of the distance.)

5-2.   If the Moon were to continue to recede from the Earth at 2 cm/year, how long would it take for the Earth–Moon system to reach double synchronous rotation? At this point, the Earth–Moon distance would be 6.0×10^5 km.

Chapter 6: Solar and Extrasolar Planets

For many decades, planetary astronomers felt frustrated by a paucity of data. While astronomers studying stars, star clusters and galaxies had countless examples of their class of objects from which to work, those trying to understand the nature and the formation of planetary systems had only one example, namely our own solar system. Now, that situation has changed. Since 1995, data on extrasolar planets has accumulated at an ever-increasing rate, and more than 300 planets have been discovered. Although it will be some time before we observe a full system like our own, certain trends are clear and have required that we revise our thinking about the formation and evolution of planetary systems.

6.1 Temperature versus Distance from a Star

It has been thought that the principal factor determining the type of planet which would occur at a given distance from a star is the temperature which a body would have at that distance due to the star's radiation falling on it. A good approximation to this temperature may be derived very simply by combining the inverse square law for radiation falling on the planet with the fourth-power law for radiation omitted by it. Let us compare two different planets at distances d_1 and d_2 from the star. The amounts of incident radiation P_{i1} and P_{i2} received by the two planets are inversely proportional to the squares of their distances from the star:

$$P_{i2} / P_{i1} = (d_1/d_2)^2 \qquad \text{(Eq. 6.1-1)}$$

For stars like the Sun, this incident energy is largely in the visible and nearby ultraviolet and infrared regions of the electromagnetic spectrum. A certain fraction of it is absorbed by the planet and warms the planet, which radiates energy in the longer wavelength infrared region. An equilibrium is reached at which the incident and emitted amounts of power are equal and the temperature remains constant. The emitted power is proportional to the fourth power of the temperature so that at equilibrium

$$P_{e1} = P_{i1} \text{ and } P_{e2} = P_{i2} \qquad \text{(Eq. 6.1-2)}$$
$$(d_1/d_2)^2 = (T_2/T_1)^4$$
$$T_2/T_1 = \sqrt{(d_1/d_2)} \qquad \text{(Eq. 6.1-3)}$$

Therefore, the equilibrium temperature of a planet which is orbiting a star and receives the bulk of its energy from the star is proportional to the inverse <u>square root</u> of its distance from the star.

Close to the star, where it is warm, ices and other volatile materials cannot condense onto planets as they are forming. The resulting planets are small, rocky and dense. Farther away from the star, where it is cold, ices and other volatile materials can condense onto planets, adding considerably to their bulk, and providing them with large enough escape velocities to retain much gaseous material as well. The resulting planets are large, gaseous, and of low density. In our own solar system, the single factor of temperature appears to explain the grouping of eight planets into four inner rocky planets and four outer gaseous ones.

6.2 Distances of Planets from a Star

After the success of Kepler's laws in describing the shapes of planetary orbits, motions along them and the relation between different orbits, several astronomers attempted to construct a theory or rule which would predict the distances of the planets from the Sun, i.e. the semimajor axes of their orbits. Kepler himself believed that these distances were related to the sizes of the five regular polyhedra (tetrahedron, cube, octahedron, dodecahedron, icosahedron) nested one inside another in a certain order. Most other astronomers correctly rejected this idea as lacking any plausible physical basis, and the search for a rule for the planetary distances continued. Finally, in 1772, Johann Bode, the director of the Berlin observatory, published the rule which became known as Bode's law, based on work done six years earlier by Johann Titius.

Bode's law is not a real physical law like Newton's laws of motion or the universal law of gravitation, but a numerical rule for predicting the distances of the planets. It can be summarized as follows:

1. Construct the series 0, 3, 6, 12, 24, etc., in which, after the 3, every number is double the preceding one.
2. Add 4 to get 4, 7, 10, 16, 28, etc.,
3. Divide by 10 to get .4, .7, 1.0, 1.6, 2.8,etc., These numbers are the predicted values of the planetary distances in AU.

The shortcoming in Bode's law was the lack of a known planet at a distance of 2.8 AU; so a search was undertaken for the missing planet. An object, albeit a small one, was found very near to the predicted distance in 1801 and named Ceres. It proved to be the largest of many asteroids whose distances generally lie in the range 2.2–3.1 AU.

Table 6.2-1 shows the actual and predicted distances for the eight major planets as well as Ceres as a typical example of the asteroids. The distances predicted by both Bode's law and the linear regression fit discussed in Chapter 1, together with their percent errors, are shown. With the exception of Neptune, the values predicted by Bode's law are remarkably accurate. The errors for the linear regression fit are somewhat larger, but this fit has the advantage that Eq. 1.2-2 is approximately equivalent to the simple power law.

$$d = .2 \times 3^{n/2} \tag{Eq. 6.2-1}$$

where n is the planetary position number.

Table 6.2-1 Actual and Predicted Distances of the Planets

Position #	Planet	Actual Distance d (AU)	Bode's Law d (AU)	Bode's Law % Error	Linear Regression d (AU)	Linear Regression % Error
1	Mercury	.387	.4	3.4	.354	−8.5
2	Venus	.723	.7	−3.2	.614	−15.1
3	Earth	1.000	1.0	0	1.06	6.0
4	Mars	1.524	1.6	5.0	1.85	21.4
5	(Ceres)	2.8	2.8	0	3.20	14.3
6	Jupiter	5.203	5.2	0	5.55	6.7
7	Saturn	9.539	10.0	4.8	9.62	.8
8	Uranus	19.19	19.6	2.1	16.7	−13.0
9	Neptune	30.06	38.8	29.1	28.9	−3.9

Currently, it is generally believed that there is probably no theory of the planetary distances which would result in a simple equation for predicting them. These distances are the result of very complex processes of planetary formation and may have changed over time. Bode's law appears to be a fortuitous result of limited applicability, although it was significant historically in that it led to the discovery of the asteroids. However, the common feature of the actual distances, as well as those predicted by both Bode's law and the linear regression fit, which probably has wider applicability, is the large spacing between adjacent planetary orbits. For example, two planets do not occur at 1.0 and 1.1 AU. There are probably good physical reasons for such well-separated orbits. If the orbits were too close, over time a collision would become nearly inevitable and only one of the planets would survive.

Based on the discussion in the previous section and this one, we would expect that although the actual values of the planetary distances in other planetary systems would be different from those in our own, these systems would have two characteristics:
1. There would be an inner group of small, rocky planets and an outer group of large gaseous planets.
2. The orbits of the planets would be well-separated.

6.3 Methods of Detecting Extrasolar Planets

From Earth-based observations, it is very difficult to observe extrasolar planets directly as points of light near their parent stars for one simple reason: they are very faint. The ratio of brightness of the star to the planet is typically a billion to one or more, and starlight scattered by the Earth's atmosphere obscures the image of the planet. Indirect detection is easier: observation of slight changes in the star's velocity, position or brightness as a result of the planet's orbiting motion. The four methods which have detected extrasolar planets summarized in Table 6.3-1.

Table 6.3-1 Methods of Detecting Extrasolar Planets

Method	Changing Quantity	Best Orientation of Orbit	Worst Orientation of Orbit	Major Strength	Major Weakness
Direct imaging	Position	Pole-on	Edge-on	Actually see planet and its motion	Planet is very faint
Doppler	Velocity	Edge-on	Pole-on	Can get lower limit on mass	No effect when viewed pole-on
Transit	Brightness	Edge-on	Pole-on	Can determine diameter	Doable with a small fraction of stars
Microlensing	Brightness			Sensitive to small-mass planets	Need very dense star field

Direct imaging of extrasolar planets has been successfully accomplished for several young stars of Spectral Class A. The Doppler method has been the most successful method so far, with more than 300 detections to its credit. The transit method yields the planet's diameter so that, if its mass is known, the density may be calculated, providing the most important clue to the planet's nature. Microlensing has produced some results recently, but requires a dense star field such as the one at the center of our galaxy.

Example Problem 6.3-1: The transit of an extrasolar planet across its parent star dims the starlight by 1%. What is the diameter of the planet relative to that of the star? Let D_s and D_p be the diameters of the star and planet, and A_s and A_p their areas, respectively.

$$\text{Brightness reduction} = A_p / A_s = 0.01$$
$$D_p / D_s = \sqrt{(A_p / A_s)}$$
$$\boxed{D_p / D_s = 0.1}$$

6.4 Results of the Extrasolar Planet Search

The great majority of the extrasolar planets detected so far share the following characteristics:

1. Masses in the range 0.5 Jupiter masses to 5 Jupiter masses.

2. Semimajor axes of less than 1 AU and in many cases of less than 0.4 AU, i.e. less than Mercury's.
3. Eccentric orbits.

The first characteristic was expected before the search began; but the second and third were not. Although, for most stars, only a single planet has been detected so far, in cases where more than one planet has been detected, they are all massive and presumably gaseous. There is no evidence so far showing the two features of the solar system stated at the end of Section 6.2.

A major effort is underway to explain the existence of massive planets close to their parent stars. The planets could not have formed where they are now due to the high temperatures. The most plausible explanation involves <u>migration</u> of the planets. In this scenario, the large, massive, gaseous planets form at large distances from the stars, where it is cold, and spiral inward due to drag of the disk of material out of which they formed, eventually settling into a much smaller orbit. Such migration might also have occurred in our own solar system but not to such a degree as to bring the Jovian planets close to the Sun. The difference may result from different amounts of material in the disk and the resultant drag. Most of the extrasolar planets we observe formed out of a massive disk in which the spiraling effect was large.

One factor that always must be borne in mind here, as elsewhere in astronomy, is the <u>bias</u> of the detection method. The Doppler method is biased toward detecting <u>massive</u> planets <u>close</u> to their parent stars, so it is not surprising that most of the planets detected are of this type. As other detection methods become available, such as direct observation of the planets from space-based observations, planets with smaller masses and/or larger distances from the star will be detected. Then, we may obtain evidence of other planetary systems which are more like our solar system.

Questions

6-1.   How does the temperature of a planet depend on its distance from the star which it orbits?

6-2.   What single factor appears to explain the grouping of the planets into two groups in our solar system?

6-3.   What is Bode's law?

6-4.   Why is there no theory of the planetary distances which results in a single equation for predicting them?

6-5.   What two characteristics would we expect that extrasolar planetary systems would share with our own?

6-6.   Why cannot extrasolar planets be seen directly by Earth-based observations?

6-7.   What four techniques have been used to detect extrasolar planets? Which has been the most successful in terms of numbers of planets?

6-8.   What two characteristics of the observed extrasolar planets were not expected before the search began?

6-9.   What is the most plausible explanation of massive planets close to their parent stars?

6-10.   What are the biases of the Doppler method?

Problems

6-1.   If the temperature of the Earth at a distance of 1 AU from the Sun is taken to be 300 K, what is the temperature of Neptune at a distance of 30 AU?

6-2.   Consider a variation of Bode's law in which the series is 0, 2, 4, 8..., the number added is 6, and the resulting series is divided by 10. What is the predicted distance of Mars and the percent error from the actual value?

6.3.   The transit of an extrasolar planet across its parent star dims the starlight by 2%. What is the diameter of the planet relative to that of the star?

Chapter 7: The Terrestrial Planets

The four inner planets of our solar system are called terrestrial because, like the Earth and unlike the four outer planets, they are relatively small, rocky and dense. However, the environments which the surfaces of the other three planets would provide to future space travelers are not at all earthlike. Earth alone has anywhere near the combination of moderate temperature, atmospheric pressure and surface gravity as well as oxygen-containing atmosphere which appear to be necessary to sustain life. Many of the differences among the terrestrial planets can be explained as a result of their different sizes and distances from the Sun, but some other factors are also important.

7.1 Mercury

Mercury is the most elusive of the five other planets conspicuously visible to the naked eye. It never strays more than 25° from the Sun and is visible only near sunrise and sunset. Most people catch a glimpse of it only rarely or even not at all in their lives. It revolves around the Sun in 87.969 days and rotates in 58.646 days so that it rotates three times during two revolutions. This 3:2 resonance, which is not synchronous rotation like the Moon exhibits, is the result of Mercury's very eccentric orbit and the large difference in the Sun's tidal forces at perihelion vs. those at aphelion.

Example Problem 7.1-1: The semimajor axis a of Mercury's orbit is 0.387 AU and its eccentricity e is 0.206. Calculate the ratio of the Sun's tidal forces at perihelion compared to those at aphelion.

The perihelion and aphelion distances, respectively, are given by
$$r_P = (1 - e)\, a$$
$$r_A = (1 + e)\, a$$
According to Eq. 5.3-5, the tidal force from the sun varies inversely as the cube of the distance so that

$$\Delta F_P / \Delta F_A = (r_A / r_P)^3$$
$$\Delta F_P / \Delta F_A = [(1 + e) / (1 - e)]^3$$
$$\Delta F_P / \Delta F_A = [1.206 / 0.794]^3$$
$$\Delta F_P / \Delta F_A = 3.5$$

The action of the Sun on Mercury may be likened to that of a juggler tossing a bowling pin. Just as the juggler grabs onto the bowling pin during the lower part of its flight and then releases it to rotate freely during the upper half, the Sun "grabs onto" Mercury near perihelion, forcing it into approximate synchronous rotation, and then lets it rotate more freely in the more distant parts of its orbit. Because Mercury has more distance to cover and moves more slowly when it is farther from the Sun, rotation gets ahead of revolution and the planet completes one and a half rotations during a revolution.

The surface of Mercury is very much like the highlands of the Moon, with a very large density of craters. The main difference from the lunar surface is the presence of scarps and wrinkles. The scarps are several-kilometer-high cliffs which extend for hundreds of kilometers. Both the scarps and wrinkles were probably formed when Mercury's surface shrank, in response to the contraction of its iron core when it solidified. Lacking an iron core, the Moon does not show the same effect. In places scarps overlay craters, indicating that the scarps are younger, whereas in others craters overlay scarps, indicating that the craters are younger. Although we do not yet have rock samples to yield accurate dates by radioactive dating, it is evident that the scarps were formed during the period of heavy bombardment by planetessimals about four billion years ago. Some of the craters are surrounded by rings of dark material and there are some volcanoes as well.

Mercury and the Moon are similar in one other respect: their rotation axes are within a few degrees of being perpendicular to the planes of their motion around the Sun (ecliptic plane for the Moon). As a result, there are areas in craters near the poles which are perpetually in shadow and where ice deposits might be located. Therefore, establishing a base on Mercury, while difficult, would not be impossible, because there would be some safe havens from the intense solar heating.

7.2 Venus

The orbit of Venus is much more circular (e = 0.007) than that of Mercury. It too does not exhibit synchronous rotation but rotates in a retrograde direction with a period of 243.16 days. The rotation is such that Venus presents the same face to the Earth at every closest approach of the two planets. It is uncertain whether this alignment is just coincidental or indicates that tidal forces from the Earth, slowed down and even reversed a more rapid initial prograde rotation.

Another factor which might be responsible for changing the rotation of Venus is the planet's dense atmosphere. Currently, the atmosphere rotates independently of the planet, also in a retrograde direction, but much more rapidly, with a period of four days. The atmosphere is massive enough that it contains a significant amount of angular momentum and could produce a slow down and reversal of the rotation. However, the amount of this effect is not clear because our understanding of the atmosphere is still incomplete.

The atmosphere of Venus consists mostly of carbon dioxide and is very dense, with a surface pressure of 95 atm, equivalent to the pressure at a depth of 1 km in the Earth's oceans. The difference in the atmospheric pressures between Venus and the Earth is striking. One key factor is the presence of water on the Earth. Much of the Earth's carbon dioxide is dissolved in the oceans, used by marine organisms and ultimately bound up in carbonate rocks. This process has kept the carbon dioxide content of the Earth's atmosphere low, whereas on Venus, most of the inventory of carbon dioxide is in gaseous form in the atmosphere.

The surface of Venus has been extensively imaged by the Magellan spacecraft, and the numerous volcanoes, lava domes and lava flows attest to widespread volcanic activity. However, unlike on the Earth, the activity is randomly distributed over the surface and not concentrated along plate boundaries. Here again, it is evidently the presence of <u>water</u> on the Earth that makes the difference. On the Earth, water which is either chemically bound to the rocks or located in the cracks between them lowers its melting point so that the rock at a depth of 100 km is partially melted and allows the tectonic plates to move. On Venus, the rocks are more solid and inhibit plate motion so that the heat escaping from the interior of the planet is released at random locations over the surface.

Because of its thick sulfuric acid cloud decks and invisible surface, radar has been important in studying Venus in several ways, and Table 7.2-1 summarizes the four major uses.

Table 7.2-1. The Four Major Uses of Radar in Studying Venus.

Measured Quantity or Feature	Technique
Distance	Distance (to planet) = speed-of-light × time
Surface elevation	Distance (to surface) = speed-of-light × time
Surface features	Brightness indicates surface roughness: dark : smooth; bright : rough
Rotation period	Doppler broadening ~ speed of rotation

For the same reason as for Venus, namely, the presence of an opaque atmosphere, radar has also been important in studying the surface of Saturn's satellite Titan. (See Section 8.4)

7.3 Mars

Mars is significantly smaller, colder and drier than the Earth and has a far lower atmospheric pressure, but it is very much like the Earth in two important respects:
1. The length of day is 24.5 hours
2. The axial tilt is 23.98° (Earth's is 23.45°)

The familiar day–night cycle and the presence of seasons will provide some earth-like comfort to any future Martian colonists, who will be faced with many other challenges in a harsh environment.

Before the era of space missions to Mars, our best views of Mars came from dedicated planetary observers who would sit at a telescope, patiently awaiting those rare moments when, for a few seconds, the turbulence in the Earth's atmosphere subsided and provided a relatively clear view. They would then attempt to memorize what they saw and then reproduce it in a sketch later. Such drawings were, in principle, superior to long exposures on photographic plates, which were dominated by periods of blurry seeing. However, they were not truly objective but were very susceptible to the subjective interpretation of the observer.

In 1877, the Italian astronomer Giovanni Schiaparelli announced that he had seen canali (Italian for 'channels') on Mars, which was widely misinterpreted as 'canals'. Following on this announcement, the American astronomer Percival Lowell built an observatory in Arizona dedicated to the observation of Mars, and in 1894 began to observe the planet over a period of years, and ultimately produced hundreds of drawings showing a network of "canals" crisscrossing the planet. He evolved a theory that explained the "canals" as constructs of a Martian civilization attempting to bring water from the polar caps to the equatorial regions of a drying planet.

Beginning in the 1960s the space missions to Mars detected canyons, dry river beds and hundreds of other features on Mars, but these are all too small to be detected from Earth in the presence of atmospheric blurring. So what was it that Schiaparelli and Lowell were seeing? The best explanation is that the "channels" or "canals" were an optical illusion, created by the human visual system struggling at the limits of perception, seeing a true difference between regions of light-colored and dark-colored soil, but straightening the boundaries and turning edges into lines.

The light-colored and dark-colored regions, with their irregular borders, can be clearly seen on images of Mars from the Mariner, Viking and other missions. However, these are superficial geologically and result from slight differences in the chemical content of the Martian soil. The only really significant surface features on Mars that can be seen from Earth are the polar caps.

We now know that water cannot exist in liquid form on the surface of Mars under the current conditions of temperature and pressure. Yet, images from many space missions provide abundance evidence of both standing and flowing water in the past. Table 7.3-1 summarizes the evidence from the four principal types of features.

Table 7.3-1. Evidence of Water on the Surface of Mars.

Feature	Probable Cause
Dry river beds	Flowing water
Terraced craters with inflow/outflow channels	Lakes and small oceans with water flowing in/out
Collapsed surface with lumpy terrain and channels	Temporary flow of water when subsurface ice was melted (by volcano?)
Craters with surrounding lobed patterns	Temporary flow of water when subsurface ice was melted by impact

The evidence is strong that Mars had a warmer and wetter climate in the past and has dried up as the atmospheric pressure has decreased, due to the loss of atmosphere because of the low escape velocity. One type of data which would help significantly in understanding the water-produced features is obtaining rock samples from which accurate dates may be determined. Obtaining such data will clearly be a goal of future space missions.

7.4 Summary of the Terrestrial Planets and the Moon

The atmospheres of the terrestrial planets and the Moon are summarized in Table 7.4-1. The bodies are listed in order of increasing escape velocity, which, in this case, is the same as increasing diameter.

Table 7.4-1. Atmosphere of the Terrestrial Planets and the Moon.

Body	Radius (km)	Density (10^3 kg/m³)	Escape Velocity (km/s)	Atmospheric Pressure (bars)	Main Constituents
Moon	1738	3.34	2.37	0	–
Mercury	2439	5.43	4.25	0	–
Mars	3397	3.93	5.04	0.007	CO_2 (95%), N_2 (2.7%)
Venus	6051	5.25	10.37	95.0	CO_2 (96%), N_2 (3.5%)
Earth	6378	5.52	11.20	1.0	N_2 (78%), O_2 (21%)

The clear rule is that, above an initial threshold, as the escape velocity increases, the amount of atmosphere increases. The notable exception is Venus; its higher temperature and consequent loss of liquid water are apparently responsible for the great density of carbon dioxide.

The surfaces of the terrestrial planets and the Moon are summarized in Table 7.4-2. The bodies are listed in order of increasing diameter.

Table 7.4-2. The Surfaces of the Terrestrial Planets and the Moon.

Body	Diameter (km)	Impact Craters	Volcanoes	Plate Tectonics
Moon	1738	Many	none	No
Mercury	2439	Many	a few	No
Mars	3397	Some	a few	No
Venus	6051	A few	many	No
Earth	6378	A few	many	Yes

The clear rule is that, as the size of the body (and its amount of residual heat) increases, the surface activity increases and reduces the number of impact craters. Although, in this case, no body is out of order, the striking difference between the presence of plate tectonics on the Earth and the lack of it on Venus cannot be explained by the small difference in size. Here again, the higher temperature of Venus and consequent loss of liquid water are apparently responsible. On Earth liquid water chemically combines with the rocks and lowers their melting points so that at a depth of 100 km in the mantle the rock is fluid enough to allow plate motion. On Venus, with no liquid water present, the melting points of the rocks are higher. The mantle rock is rigid and no plate motion can occur.

Questions

7-1. Why does Mercury exhibit a 3:2 resonance rather than synchronous rotation?

7-2. How do we know that the scarps on Mercury were formed during the period of heavy bombardment?

7-3. What might explain the slow retrograde rotation of Venus?

7-4. What is the likely explanation of the lack of plate tectonics on Venus?

7-5. What are the four major uses of radar in studying Venus?

7-6. In what two ways is Mars similar to Earth?

7-7. How did observers of Mars in the late nineteenth and twentieth centuries attempt to create a clear image of the planet?

7-8. What is the best explanation of the "channels" or "canals" on Mars seen by Schiaparelli and Lowell?

7-9. What features provide evidence of water on Mars?

7-10. Why is the atmosphere of Venus so much denser than would be expected from its escape velocity?

7-11. How do surface activity and the number of observed impact craters depend on size for the terrestrial planets and the Moon?

Problems

7-1. When it is closest to the Earth, Venus is ~40 million km away. How long does it take a radar signal to go out to Venus and return?

Chapter 8: The Jovian Planets

The giant gaseous planets of the outer solar system, also called the jovian planets as they resemble Jupiter, are qualitatively different from the four terrestrial planets. They are much larger, much less dense and all have extensive systems of satellites and rings. It is believed that they formed in the outer cooler parts of the solar nebula where the ices remained solid and so could grow to much larger size. There was also abundant material with which to make satellite systems by accretion on a much smaller scale than for the entire solar system.

8.1 Physical Properties

The main physical properties of the jovian planets are listed in Table 8.1-1. There are a number of expected trends in the values, but there are also some unusual values (underlined) which call for special comment.

Table 8.1-1. Physical Properties of the Jovian Planets.

Physical Property	Jupiter	Saturn	Uranus	Neptune
Equatorial radius (km)	71,492	60,268	25,559	24,764
Polar radius (km)	66,850	54,362	24,974	24,343
Oblateness $(r_e - r_p)/r_p$	.069	.098	.023	.017
Mass (kg)	1.90×10^{27}	5.69×10^{26}	8.68×10^{25}	1.02×10^2
Mass (relative to Earth)	318.	95.2	14.5	17.1
Average density (10^3 kg/m³)	1.33	0.69	1.29	1.64
Gravity (Relative to Earth)	2.4	.93	.79	1.12
Escape velocity (km/s)	59.6	35.5	21.3	23.3
Interior rotational period (hr)	9.925	10.675	17.233	16.117
Inclination of rotation axis (°)	3.1	26.7	97.9**	29.6
Magnetic field strength (relative to Earth)	14	.71	.74	.43
Inclination of magnetic axis (°)	9.6	0	58.6	46.8
Offset from center (fraction of the radius)	0	0	.30	.53
Interior layers (as a fraction of equatorial radius)				
Core	rock&ice .18	rock&ice .27	rock .29	rock .28
Mantle	liq.met H .52	liq.met. H .25	water? .68	water? .69
	liq.mol H .25	liq.mol. H .46		
Atmosphere	gases .02	gases .02	gases .03	gases .03
atmospheric composition (fraction)				
Hydrogen	.955	.97	.90	.86
Helium	.044	.015	.09	.12
Methane (CH4)	.0069	.001	.01	.00
Ammonia (NH3)	.0002	.001	-	-
Atmospheric activity	Very strong	Moderately strong	Weak	Strong

* At the height of the cloud deck.
** Retrograde rotation.

Sizes and masses generally decrease with distance from the Sun, probably due to the thinning out of the solar nebula with distance. Because of its similar composition and smaller mass (and resulting gravitational compression), Saturn is less dense than Jupiter. However, the densities increase from Saturn through Neptune, owing to the greater proportion of heavier elements in the outer planets.

All four planets rotate rapidly and are noticeably oblate, i.e. flattened at the poles. Although Saturn rotates a bit slower than Jupiter, its much lower gravity accounts for greater oblateness. The unusual rotation is that of Uranus, which rotates in a retrograde direction with its axis lying nearly in its orbital plane. The most likely cause appears to be a giant impact which knocked the planet on its side when it was young.

Because of the combination of rapid rotation and liquid interiors (metallic hydrogen in Jupiter and Saturn, and probably water in Uranus and Neptune), all four planets have significant magnetic fields. Jupiter's is by far the most extensive, and the plentiful source of particles emitted by Io's volcanic eruptions creates a huge and lethal radiation belt near Jupiter. Only at the distance of Callisto would the radiation fall to non-lethal levels and even there prolonged exposure would still be undesirable. The magnetic fields of Uranus and Neptune are unusual in that their magnetic axes are highly inclined and significantly offset from the center of the planet.

The trends in atmospheric composition are straightforward with the exception of the very low helium content on Saturn. For reasons not fully understood, much of the helium on Saturn has sunk into the interior. Atmospheric activity, driven mostly by internal heat at these great distances from the Sun, would be expected to decrease with planetary mass. The atmosphere of Uranus is less active than expected while that of Neptune is more so. The reason for this striking difference is also not fully understood.

8.2 The Roche Limit

In 1849, the French scientist M. E. Roche showed that, within a certain distance of a planet, the planet's tidal forces on a satellite would exceed the satellite's own self gravitational forces and a satellite held together by gravity could not exist. In order to show that such is the case, we must consider the balance of tidal and self-gravitational forces depicted in Fig. 8.2-1.

Two small bodies, assumed for simplicity to be identical spheres of mass (M_s), radius (R_s), and density (ρ_s), orbit a planet of mass, M_p, radius, R_p, and density, ρ_p at a distance of r. The planet's gravitational force on the closer body is stronger and the differential tidal force tending to pull the two bodies apart can be written, using Eq. 5.3-5 and the fact that $\Delta r = 2R_s$ as

$$\Delta F = 4GMmR_s/r^3 \qquad\qquad (Eq.\ 8.2\text{-}1)$$

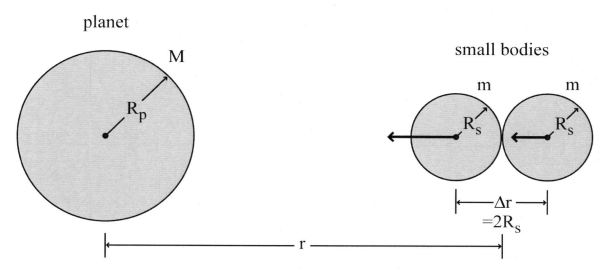

Fig. 8.2-1. Tidal and Self-Gravitational Forces in the Vicinity of a Planet.

The attractive gravitational force between the two small bodies is given by

$$F = Gmm/(\Delta r)^2$$
$$F = Gm^2/(4R_s^2) \qquad \text{(Eq. 8.2-2)}$$

Far from the planet, the self-gravitational force exceeds the tidal force and the small bodies will be pulled together into a larger one. Close to the planet, the tidal force is larger and the two bodies will be pulled apart. These forces will be equal at the Roche distance r_R which is determined by equating the two expressions in Eqs. 8.2-1 and 8.2-2. Using the fact that $M = 4\pi\rho_p R_p^3/3$ and $m = 4\pi\rho_s R_s^3/3$ we may express r_R in terms of the planetary radius R_p and the densities ρ_p and ρ_s:

$$r_R = (16\rho_p/\rho_s)^{1/3} R_p \qquad \text{(Eq. 8.2-3)}$$

If the densities of the planet and the small bodies are comparable, then $r_R = 2.5 R_p$.

Example Problem 8.2-1: Using $\rho_p = .69$ and $\rho_s = 1.12$, calculate the Roche distance for Saturn in terms of planetary radius.

$$r_R = (16 \times .69/1.12)^{1/3} R_p$$

$$\boxed{r_R = 2.14 R_p}$$

8.3 Rings

Although only Saturn has rings visible from Earth, all four Jovian planets have ring systems, whose main properties are summarized in Table 8.3-1.

Table 8.3-1

Planet	Jupiter	Saturn	Uranus	Neptune
Number of Rings Narrow Wide	3 0 3	Many Many Many	10 8 2	4 2 2
Thickness	<30 km for main	<100 km for 3 main	≤ 100 m	
Particle Size (m)	Dust	.01 to 10	.2 to 5	
Particle Brightness	Dark	Bright	Dark	Dark
Shepherd Satellites		F ring	E Ring	Inner narrow ring
Cleaning Satellites		Encke division by 1981 S13		
Resonance Satellites		Cassini division with Mimas		
Source of Material	Dust knocked off Adrastea	Comets? stray satellites	Comets?	Comets?

The ring systems share several common properties. The extreme thinness of most of the rings of Saturn and Uranus is maintained by frequent collisions in the dense ring plane which prevent a ring particle which has been knocked into an inclined orbit from maintaining that inclination for very long. The Saturnian and Uranian systems also have similar ranges of particle size. All of the ring systems contain dust.

The presence of shepherd satellites, orbiting near the inner and outer edges of narrow rings and preventing them from spreading out has been confirmed in these cases and is suspected to be the case for all narrow rings. Jupiter's ring system is the one in which the source of material is known: it is dust knocked off the nearby small satellite Adrastea by micrometeorite bombardment. It is thought that the other ring systems must also have a source of material, perhaps comet nuclei or stray satellites, because the ring particles continue to get smaller with successive collisions and ultimately spiral into the planet's atmosphere in a time much shorter than the age of the solar system.

Saturn's ring system is by far the most extensive and complex, and exhibits several unique features: (1) There are gaps in the rings produced by a resonance effect (particles in the Cassini division have an orbital period which is exactly half that of the satellite Mimas) and a cleaning satellite (the small satellite 1981 S13 sweeps the Encke division clean). (2)

There are radial spokes in the ring system probably caused by the interaction of electrically charged particles with Saturn's rapidly rotating magnetic field. (3) Bunching of the ring particles and density waves can also be seen.

The study of the ring systems of the Jovian planets is fascinating in itself, but also has much broader applications. Disk-shaped structures are important in larger context, as in the formation of planetary systems and in spiral galaxies. The ring systems of our own solar system provide an opportunity to study the dynamics of disk-shaped systems from a relatively close vantage point and in real time.

8.4 Satellites

The principle satellites of the Jovian planets comprise six large satellites (radius = 1350 – 2700 km) and thirteen medium-sized ones (radius = 170 – 800 km). In addition, each planet has many other small satellites. The orbital properties of the principle satellites are listed in Table 8.4-1.

Table 8.4-1. Orbital Properties of the Principal Satellites of the Jovian Planets.

Planet	Satellite	Distance (10^6 km)	Period (days)	Inclination (°)	Eccentricity
Jupiter	Io	0.4216	1.769	0.04	0.00
	Europa	0.6709	3.551	0.47	0.01
	Ganymede	1.0700	7.155	0.19	0.00
	Callisto	1.8830	16.689	0.28	0.01
Saturn	Mimas	0.18552	0.942	1.53	0.02
	Enceladus	0.23802	1.370	0.02	0.00
	Tethys	0.29486	1.888	1.09	0.00
	Dione	0.3774	2.737	0.02	0.00
	Rhea	0.52704	4.518	0.35	0.00
	Titan	1.22185	15.945	0.33	0.03
	Iapetus	3.5613	79.331	14.7	0.03
Uranus	Miranda	0.1298	1.413	3.4	0.0
	Ariel	0.1912	2.520	0.0	0.0
	Umbriel	0.2660	4.144	0.0	0.0
	Titania	0.4358	8.706	0.0	0.0
	Oberon	0.5862	13.463	0.0	0.0
Neptune	Proteus	0.1776	1.211	0.0	0.0
	Triton	0.35459	5.875	157.*	0.0
	Nereid	5.5886	360.125	29.	0.75

*Retrograde motion.

Except for Iapetus and Nereid, the orbital periods range from about one day to a little over two weeks. For the satellites which have orbital periods with this range, the orbits all have a low angle of inclination relative to the planetary equator and a very low eccentricity. The large equatorial bulges of the Jovian planets act to pull the satellites into equatorial orbits and the tidal bulges raised by the satellites on the planets act to circularize the satellites' orbits. The only exception is Triton, which revolves in a retrograde direction around Neptune. This exceptional orbit, unique among the principle satellites, has prompted speculation that Triton is a captured body, since it is difficult to imagine how the accretion process could create a satellite orbiting in a retrograde fashion.

The physical properties of the principal satellites are listed in Table 8.4-2. Of particular interest are the trends in density and the relative activity of the surfaces.

Table 8.4-2. Physical Properties of the Principal Satellites of the Jovian Planets.

Planet	Satellite	Radius (km)	Mass (10^{21} kg)	Density (10^3 kg/m^3)	Notable Features
Jupiter	Io	1821	89.33	3.57	Many sulfur volcanoes, no craters
	Europa	1565	47.97	2.97	Icy surface with cracks, few craters
	Ganymede	2634	148.2	1.94	Craters and grooved terrain
	Callisto	2403	107.6	1.86	Old cratered surface, multi-ringed basin
Saturn	Mimas	200	0.0370	1.12	Giant impact crater-bright icy surface, geysers
	Enceladus	250	0.06	1.00	Cratered terrain, long wide valley
	Tethys	530	3.617	0.98	Crated, wispy terrain
	Dione	559	1.08	1.49	Smooth, wispy terrain, thick
	Rhea	764	2.31	1.24	
	Titan	2,575	134.55	1.88	N$_2$ atmosphere, channels methane/ethane lakes, craters
	Iapetus	720	1.59	1.00	Bright/dark surface
Uranus	Miranda	235	.066	1.26	Mixed terrain, high cliffs
	Ariel	580	1.35	1.65	
	Umbriel	585	1.17	1.44	Old crated surface
	Titania	789	3.52	1.59	Younger icy surface
	Oberon	761	3.01	1.50	Old cratered surface
Neptune	Proteus	210	.06	1.55	
	Triton	1350	21.4	2.14	Cantaloupe terrain, N$_2$ geysers, thin atmosphere
	Nereid	170	.031	1.51	

The satellite system of Jupiter is the only one with a density sequence, i.e. decreasing density with increasing distance from the planet. Apparently, Jupiter, like the Sun, was hot enough during its formation to prevent significant amounts of ices from condensing on Io and Europa so that they consist mainly of rock. All the other satellites listed have much lower densities and are mixtures of rock and ice. The average densities of the medium-sized satellites have a minimum at Saturn (1.14) and then increase at Uranus (1.49) and Neptune (1.53). It appears that the water ice content in the outer solar system is a maximum at the position of Saturn, a fact which is reflected in both the low density of its satellites and the extent of its ring system.

As with solid planets, it would be expected that surface activity should correlate with size, but there are many exceptions among the jovian satellites. In three cases (Io with Europa, Europa with Ganymede and Enceladus with Dione), a satellite is caught in a 2:1 resonance with an outer satellite whose orbital period is twice as long. The outer satellite tends to pull the inner one into a more elliptical orbit while the parent planet tends to pull it into a more circular one, causing tidal flexion and an input of heat. This tidal heating results in sulfur volcanoes on Io, ice cracks and ridges on Europa, and ice ridges and H_2O geysers on Enceladus. Titan is active because of its thick atmosphere and is the only other body in the solar system besides the Earth whose surface has standing and flowing liquid, presumably methane and/or ethane in the case of Titan.

Questions

8-1.   Why do the densities of the outer planets increase from Saturn through Neptune?

8-2.   What is the most likely cause of the unusual orientation of the rotation axis of Uranus?

8-3.   What is the Roche limit?

8-4.   If the densities of a planet and its small bodies are comparable, what is the Roche limit in terms of planetary radii?

8-5.   Name three common properties shared by at least two of the ring systems.

8-6.   What are the broader applications of the study of planetary ring systems?

8-7.   Why do most of the satellites of the jovian planets have circular equatorial orbits?

8-8.   What is exceptional about the orbit of Triton? What does it suggest about Triton's origin?

8-9.   Which is the only satellite system with a density sequence?

8-10. Which satellites have unusual surface activity? Why?

Problems

8-1.   Using $\rho_p = 1.29$ and $\rho_s = 1.50$, calculate the Roche limit for Uranus in terms of planetary radii.

Chapter 9: Dwarf Planets and Small Bodies of the Solar System

The smaller bodies of the solar system are not as easy to classify as are the major planets, which divide neatly into the four inner terrestrial planets and four outer jovian planets. Their sizes, densities, and orbits are much more varied, and they are far harder to detect and observe. After much thought and deliberation, astronomers have decided to group these objects into dwarf planets, of which there are presently three, but ultimately probably more, and small bodies of the solar system, of which there are countless trillions. Although the combined mass of the small bodies is far less than that of the major planets, their large numbers render them very important. Impacts by these objects produced the many cratered surfaces of the solid planets and satellites and may have altered the course of life on Earth.

9.1 Dwarf Planets

In order to qualify as a dwarf planet, a celestial object must satisfy the following four conditions:
1. It must be in orbit around the Sun.
2. It must have sufficient gravity for its own mass to have pulled it into a roughly spherical shape.
3. It has not cleared the neighborhood around its orbit.
4. It is not a satellite.

At present, there are three objects which surely meet these criteria, and their characteristics are given in Table 9.1-1.

Table 9.1-1. Characteristics of Dwarf Planets.

Name	Ceres	Pluto	Eris
Location	asteroid belt	Kuiper belt	scattered disk
Year Discovered	1801	1930	2005
Diameter (km)	940	2310	2400
Mass (10^{21}kg)	.95	13.05	16.0
Density (10^3kg/m^3)	2.08	2.0	2.1
Semimajor Axis (AU)	2.766	34.482	57.68
Period (yr)	4.599	248.09	557.
Eccentricity	.090	.248	.442
Inclination (°)	10.587	17.142	44.187
Number of Satellites	0	3	1

There are 10–20 other bodies with diameters greater than 750 km which may also qualify as dwarf planets, but require further study. These include several of the largest asteroids as well as the recently discovered objects Orcus, Sedna and Quaoar in the Kuiper belt and scattered disk.

One of the potential dwarf planets is Pluto's satellite Charon, which has a diameter of 1260 km, a mass of 1.77×10^{21} kg, a density of 1.8×10^3 kg/m^3, and orbits Pluto at a distance of 19,600 km in a period of 6.38718 days. If Charon qualifies, then probably it is best to classify the Pluto–Charon system as a double dwarf planet, because the center of mass, i.e. barycenter, lies outside of Pluto, in space, and distinguishes this system from the Earth–Moon system and all other planet–satellite systems.

9.2 Asteroids

Most asteroids stay in the main asteroid belt between Mars and Jupiter. Although it appears that they almost touch each other in typical plots of the asteroid belt, the apparent large density results from not being able to depict accurately their true small size. Their average separation is several million kilometers, and, with the exception of the few such as Sylvia which have satellites (see Problem 9.2), a visitor would be unable to glimpse an asteroid's neighbor with the naked eye. Orbital resonances with the planet Jupiter, in which the asteroids' period is to Jupiter's period (~12 yr) as a ratio of smaller integers, e.g. 1:2, 1:3, 1:4, 2:5, are responsible for gaps in the belt known as Kirkwood gaps. It is believed that, during the formation of the solar system, this effect prevented a sufficient amount of material from accreting to form a major planet.

The two major types of asteroids are rocky asteroids with a density of 2.3×10^3 kg/m^3 and metallic asteroids with a density of $\sim 7 \times 10^3$ kg/m^3. In the case of the rubble-pile asteroids, loose aggregates of rock with much space in between, the density may be considerably lower. The most plausible explanation for the two distinct groups is that the accretion process proceeded to the point where the growing body was large enough to undergo differentiation, and then a collision broke it apart. Fragments of the core become metallic asteroids and fragments of the mantle become rocky ones. Someday, we may be able to benefit from Nature's refining process and make use of the nearly pure metallic material of the metallic asteroids. The energy required to lift material off the surface will be small due to the low gravity.

A very small fraction of asteroids have orbits which cross the Earth's, and there is a small probability that someday one might strike the Earth. Table 9.2-1 shows for three ranges of size a rough estimate of the numbers of asteroids in the range, together with the kinetic energy and the consequences of impact.

Table 9.2-1. Effect of Asteroid Impacts on the Earth.

Size range (km)	Number	Energy (10^6 tons of TNT)	Effects
.1	Several $\times$ 10^5	75	Severe regional destruction
1	Several $\times$ 10^3	75,000	Severe global destruction
10	Several	75,000,000	Global catastrophe, extinction of species

Because the kinetic energy of the impacting asteroid is equal to $mv^2/2$, and the speed of impact is typically in the range 40–50 km/s, the main energy dependence is proportional to the mass, which varies as the <u>cube</u> of the diameter. Although the probability of a severe impact is low, the consequences would be so devastating that prevention is essential. Prevention consists of <u>early detection</u> followed by <u>deflection</u>, perhaps in the form of landing a rocket propulsion system on the asteroid and giving it a steady gentle push during several months. The difference in its trajectory from the slightly altered velocity would accumulate over the years and cause it to miss the Earth. Tracking of Earth-crossing asteroids must be improved, particularly for the lowest size range, where the numbers are highest.

9.3 Kuiper Belt and Scattered Disk Objects

The Kuiper belt and the scattered disk are located beyond the orbit of Neptune and contain millions of small bodies. The first Kuiper belt object was discovered in 1992 and now hundreds are known. These small bodies are composed mainly of ice and rock and are usually coated with a dark material which probably contains carbon. Their dark surfaces and great distances render them difficult to detect. Undoubtedly, other large objects, some even larger than Pluto, await discovery.

9.4 Comet Nuclei

Most comet nuclei reside in a very extended halo around the solar system known as the Oort cloud. In the mid twentieth century, Jan Oort calculated the aphelion distances of comets with very eccentric orbits and concluded that they were coming from tens of thousands of astronomical units away. He realized that such a large reservoir of comet material is necessary to replenish the supply of periodic comets such as Halley's which lose ~1/100000 of their mass on each passage of the Sun and last but a small fraction of the age of the solar system. The creation of a new periodic comet involves two steps: (1) deflection into the inner solar system by a passing star or other interstellar body and (2) capture in the inner solar system by Jupiter or one of the other jovian planets.

Comet nuclei range in size from 1 to 10 km and consist of porous ice with a typical density of ~.3 $\times$ 10^3 kg/m^3. They lack sufficient self gravity to compact themselves into truly solid masses. Fred Whipple coined the term "dirty snowball" to describe them. The flybys of

Halley's comet in 1986 showed that the evaporation of material off the comet nucleus occurs in the form of <u>jets</u> which presumably form, run their course, and then yield to others. This localized evaporation creates the coma and tail and produces the sometimes spectacular phenomenon of a comet.

9.5 Annual Meteor Showers and Meteor Storms

Much of the larger bits of material evaporated off the comet nucleus remain in the comet's orbit, gradually spreading away from the comet nucleus over the years. Some comets' orbits intersect the Earth's orbit and when the Earth passes through the comet's orbit every year, it encounters the stream of residual material. Small bits of ice, typically the size of a penny, burn up as they enter the Earth's atmosphere and create an <u>annual meteor shower</u>. Because the material exists in the form of braids and filaments which are moved around by the Earth and the other planets, the count rate of a meteor shower is hard to predict. During the best years, the most reliable shower, the Perseids, which usually occurs on the night of August 11–12, can yield a count of several hundred per hour.

On rare occasions, the count rate may be ~100 times higher still during a <u>meteor storm</u>. There are two conditions for such an event: (1) the Earth has to pass through the comet's orbit and (2) the comet nucleus and associated denser stream of debris has to be near the Earth. Not all comets' orbits have such a denser stream or still contain a comet nucleus. One which does is the orbit of comet 55P Tempel-Tuttle, whose nucleus has a period of 33.33 years. This orbit produces the annual shower known as the Leonids, which occurs on the night of November 16–17. Every 33 +/– 2 years more intense showers usually occur and in some years, namely 1799, 1833, and 1966 there have been spectacular meteor storms with count rates of tens of thousands per hour. Because the streams of material producing these events are so narrow, only selected locations on Earth are able to witness them and accurate prediction is not yet possible.

Questions

9-1. What are the four criteria for a dwarf planet?
9-2. Why might the Pluto–Charon system legitimately be considered a double dwarf planet?
9-3. Why do some asteroids have a density less than that of water?
9-4. What are the two steps necessary for prevention of a catastrophic asteroid impact on the Earth?
9-5. Why are Kuiper belt and scattered disk objects hard to detect?
9-6. If the Oort cloud did not exist, what would eventually happen to periodic comets?
9-7. What is the typical composition and size of the particles producing a meteor shower?
9-8. Explain the difference between an annual meteor shower and a meteor storm.

Problems

9-1. Calculate the angular size of Charon (diameter = 1260 km) as seen from Pluto (diameter 2310 km) and vice versa. The Pluto–Charon distance is 19,600 km.
9-2. Calculate the angular size of the satellite asteroids Romulus (diameter = 18 km) and Remus (diameter = 7 km) as seen from Sylvia. Romulus is at distance of 1360 km and Remus is at a distance of 710 km.
9-3. Calculate the mass of Sylvia from Kepler's third law using: (a = 1360 km, P = 87.6 hr) for Romulus and (a = 710 km, P = 33 hr) for Remus. Then calculate the density of Sylvia by approximating it as a sphere of radius 140 km.
9-4. Calculate the energy E required to lift one ton (10^3 kg) of metal off a metallic asteroid with a density of 7×10^3 kg/m^3 and a diameter of 40 km. Use the relation E = GMm/R where M and R are the mass and radius of the asteroid and m = 1 ton.
9-5. Calculate the lifetime of Halley's Comet, given that it travels near the Sun every 76 years and loses 1/100000 of its mass as it does so.

Chapter 10: Electromagnetic Radiation and Spectra

Most of our knowledge of the universe beyond our own solar system comes from an analysis of the electromagnetic radiation emitted by celestial objects. For nebulae, star clusters and galaxies, we see an image, but stars present such a small angular size that they are essentially points of light. Spatial features of stars must be inferred from features in their spectra using basic physics and plausible models. In the last half century, observational astronomy has expanded from the visible and radio regions of the electromagnetic spectrum into the ultraviolet, X-ray and gamma-ray regions at short wavelengths and the infrared, far-infrared and microwave regions at long wavelengths. For each object studied, each region of the spectrum provides a piece of the entire intricate and complex picture.

10.1 Blackbody Radiation: The Blackbody Spectrum

The standard spectrum against which stellar and other spectra may be compared is the blackbody spectrum. As defined in physics, a blackbody is an object which is at a uniform temperature T (in Kelvins), omits all of its electromagnetic radiation because of thermal energy associated with its temperature, and reflects no other radiation. Such an ideal object does not exist in nature but can be approximated by an enclosed cavity with a small hole in it maintained at temperature T. Any radiation incident into the hole will be absorbed and its energy re-emitted back out through the hole in the cavity's own thermal spectrum. At first, it might appear that a star does not resemble such a cavity at all, but consider the important criterion of zero reflection. Imagine shining a laser beam at a star. Just as in the case of the cavity, the laser light will not be reflected but will be gradually absorbed by the stellar atmosphere and its energy re-emitted in the star's own thermal spectrum. For stars, the main departure from the blackbody ideal results from the fact that its absorbing and emitting regions are not all at a uniform temperature T.

Experiments in the late nineteenth century established two important properties of the blackbody spectrum. First, the peak wavelength, i.e. the wavelength at which the blackbody emits the most power, is inversely proportional to its temperature T:

$$\lambda_{max} T = \text{constant} \qquad \text{(Eq. 10.1-1)}$$

This relation is known as <u>Wien's law</u> and the value of the constant depends on the way the spectrum is plotted (See discussion below). Second, the total flux, i.e. power per unit surface area, emitted by a blackbody is proportional to the fourth power of its temperature T:

$$F = \sigma T^4 \qquad \text{(Eq. 10.1-2)}$$

This relation is known as the Stefan–Boltzmann law and the Stefan–Boltzmann constant σ has the value 5.6705×10^{-8} W/m²/K⁴. In 1900, the German physicist Max Planck introduced the concept of quanta of radiation, called photons, and derived an expression for the blackbody spectrum which is given in normalized form on a logarithmic wavelength scale as

$$f = (15 \times \ln10/\pi^4)x^4/(e^x-1) \qquad \text{(Eq. 10.1-3)}$$

where the dimensionless quantity $x = hc/\lambda kT$ is the ratio of the photon energy hc/λ to the thermal energy kT. The three constants appearing in this ratio are:

h = Planck's constant = 6.62608×10^{-34} Js
c = speed of light = 2.997925×10^{8} m/s
k = Boltzmann's constant = 1.380×10^{-23} J/K

The normalized blackbody spectrum on a logarithmic wavelength scale is shown in Fig. 10.1-1. On this plot, the values of log λ and λ on the abscissa are relative to the peak wavelength λ_{max}. The value of the flux f (power/area/decade) on the ordinate is relative to the total or integrated flux F (power/area). (Remember that 'decade' means an increment of one unit in log λ and a factor of ten in λ.) Because the flux has been normalized, the total area under the curve is unity. Fig. 10.1-1 is also provided as a transparency which

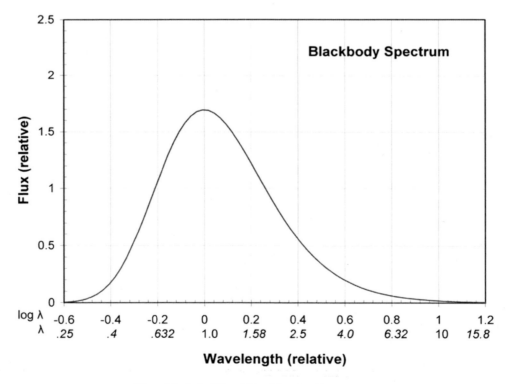

Fig. 10.1-1. The Blackbody Spectrum.

can be placed over a stellar spectrum and slid from side to side to achieve a best fit by eye and obtain the corresponding blackbody temperature. Note that the blackbody spectrum is asymmetric, with a longer tail on the long-wavelength side than the short-wavelength side.

Example Problem: Suppose a star has a blackbody spectrum with a peak wavelength ($\log \lambda_{max} = 2.7$, $\lambda_{max} = 500$ nm) and a total flux F = 100 nW/m² (1 nW = 1 nanowatt = 10^{-9} W). If the transparency of the blackbody spectrum is placed over the stellar spectrum, what do the relative values $\log \lambda = 0, -0.3, 0.3$ and f = 1.0, 1.7 correspond to in absolute units.

Relative log λ	Absolute log λ	Relative λ	Absolute λ (nm)	Relative f	Absolute f (nW/m²/dec)
0	2.7	1.0	500	1.0	100.0
−0.3	2.4	0.5	250	1.7	170
0.3	3.0	2.0	1000		

The Logarithmic Wavelength Scale

Astronomers working in different parts of the electromagnetic spectrum usually plot spectra in their own set of units. Ultraviolet, visible and infrared astronomers plot flux per wavelength interval, f_λ, in units such as W/m²/Å (watts per meter squared per Angstrom). Radio astronomers plot flux per frequency interval, f_ν, in janskys (10^{-26} W/m²/Hz), i.e. 10^{-26} watts per meter squared per hertz. X-ray astronomers may use either f_λ or f_ν. This lack of unity can be very confusing. In this book, for all spectra, in all regions of the electromagnetic spectrum, we shall plot flux per logarithmic wavelength (or, equivalently, frequency) interval in units of W/m²/dec (watts per meter squared per decade) or units using one of the following multiples of a watt:

nW	nanowatt	10^{-9} watt
pW	picowatt	10^{-12} watt
fW	femtowatt	10^{-15} watt

There are four principal advantages to using the logarithmic wavelength scale:
1. A plot can show the spectrum over a large dynamic range of wavelength (or, equivalently, frequency or energy).
2. The flux and total or integrated flux are of comparable magnitude.
3. The blackbody spectrum keeps the same shape as the temperature varies. Only its position along the wavelength axis changes.
4. The Doppler effect (See Section 10.3) displaces the entire spectrum by the same amount.

The second advantage deserves some comment. Note in Fig. 10.1-1 that for the ideal blackbody spectrum, the peak flux level (e.g., in nW/m²/dec) is almost exactly 1.7 times the total flux (e.g., in nW/m^2). This is an easy number to remember when comparing the blackbody spectrum with stellar spectra which have been normalized. If the peak in the stellar spectrum is narrower than that of the blackbody spectrum, the ratio will be greater than 1.7 times, whereas if the peak is broader, the ratio will be less, because the area under the curve is unity. Departures from the blackbody ideal are immediately obvious.

Caution: In using the logarithmic wavelength scale, be careful to use the correct constant in Wien's law as given in table 10.1-1 below.

Table 10.1-1. Constants in Wien's Law for f_λ, f, f_v.

Quantity	Scale	Constant (nmK)
f_λ	Linear wavelength	2.898×10^6
f	Logarithmic wavelength or frequency	3.6715×10^6
f_v	Linear frequency	5.1020×10^6

Relative to the logarithmic scale, the linear wavelength scale tends to stretch the spectrum at long wavelengths and so reduce λ_{max}, while the linear frequency scale does just the opposite. For the logarithmic scale, Wien's law may be expressed logarithmically as

$$\log \lambda_{max} + \log T = 6.565 \qquad \text{(Eq. 10.1-4)}$$

Flux of Radiation from a Blackbody

Let us represent a star by a spherical blackbody of radius R and temperature T. The star is observed from a distance r, where $r \gg R$. The surface area of the star is $4\pi R^2$ and the power emitted per unit area is σT^4 so that the total power radiated by the star, i.e. its luminosity is given by

$$L = 4\pi R^2 \sigma T^4 \qquad \text{(Eq. 10.1-5)}$$

At a distance r from the star, the total luminosity is distributed uniformly over a spherical surface of area $4\pi r^2$ so that the observed flux (power/area) is

$$F = L / 4\pi r^2$$

$$F = (R/r)^2 \sigma T^4 \qquad \text{(Eq. 10.1-6)}$$

The angular radius α of the star is R/r and the corresponding solid angle Ω is $\pi \alpha^2 = \pi(R/r)^2$ so that the flux per solid angle is given by

$$F/\Omega = \sigma T^4 / \pi \qquad \text{(Eq. 10.1-7)}$$

Solving for T we obtain

$$T = (\pi F / \sigma \Omega)^{1/4} \qquad \text{(Eq. 10.1-8)}$$

This temperature is known as the <u>effective temperature</u> and is the temperature of a blackbody which, having the same angular radius and solid angle as the star, would produce the same total flux. The temperature mentioned above, determined by matching the position along the wavelength scale of a normalized blackbody spectrum to the normalized stellar spectrum, is known as the <u>color temperature</u>. Usually, because absorption lines and bands tend to rob power from the stellar spectrum and reduce the flux, the effective temperature is less than the color temperature. In the exercises in this book, we shall be concerned with the color temperature only.

10.2 Absorption and Emission Line Spectra

In real stellar spectra, there are wavelengths at which the flux is considerably less or greater than the (approximately blackbody) continuum level. Such absorption and emission lines are caused by atoms, ionized atoms or molecules in the stellar atmosphere which absorb or emit photons at certain discrete wavelengths. The simplest atom is hydrogen, for which there is a simple formula for the allowed wavelengths. All other atoms, even helium, are far more complicated. In a hydrogen atom, the energy levels of an electron are given by

$$E_n = -E_o / n^2 \qquad\qquad \text{(Eq. 10.2-1)}$$

where $E_o = 2.1793 \times 10^{-18}$ J, n is an integer denoting the level, and the negative sign indicates an energy of binding to the proton. Absorption of a photon is accompanied by a change from a lower level n_1 to a higher level n_2 and emission of a photon by change from a higher level n_2 to a lower level n_1. In either case, the energy of the photon hc/λ is equal to the difference in energy between the levels:

$$hc/\lambda = E_o \, (1/n_1^2 - 1/n_2^2) \qquad\qquad \text{(Eq. 10.2-2)}$$

so that the wavelength is given by

$$\lambda/\lambda_o = 1 / (1/n_1^2 - 1/n_2^2) \qquad\qquad \text{(Eq. 10.2-3)}$$

where $\lambda_o = 91.15$ nm and is called the Lyman limit.

The two most important series of hydrogen lines in stellar spectra are the Lyman series ($n_1 = 1$), which occurs in the far ultraviolet and the Balmer series ($n_1 = 2$), which occurs in the visible and near ultraviolet. The lines are labeled by the letters of the Greek alphabet and their relative and absolute wavelengths and their displacement on a logarithmic wavelength scale relative to the limiting wavelength of the series ($n_2 = \infty$) are given in Table 10.2-1.

Table 10.2-1. The Lyman and Balmer Series of Hydrogen.

The Lyman Series ($n_1 = 1$)					The Balmer Series ($n_1 = 2$)					
Letter	n_2	λ/λ_o	λ (nm)	log (λ/λ_o)	Letter	n_2	λ/λ_o	λ (nm)	log (λ/λ_o)	Δlog (λ/λ_o)
α	2	4/3	121.53	.125	α	3	36/5	656.3	.857	.256
β	3	9/8	102.54	.051	β	4	16/3	486.1	.727	.125
γ	4	16/15	97.23	.028	γ	5	100/21	434.0	.678	.076
δ	5	25/24	94.95	.018	δ	6	36/8	410.2	.653	.051
∞	∞	1	91.15	0	∞	∞	4	364.6	.602	0

Look carefully at the final column for each series. Note that the spread of the Balmer series is more than twice that of the Lyman series on a logarithmic wavelength scale. Since any Doppler shift preserves these relative displacements, there is no way that a strongly red-shifted Lyman series can masquerade as a Balmer series. (See Section 16.2 for further discussion.)

Figure 10.2-1 shows how absorption and emission lines are produced in a star. The continuous, approximately blackbody spectrum is created in the visible surface of the star, which is called the photosphere. If the gas at the appropriate height for the given atomic,

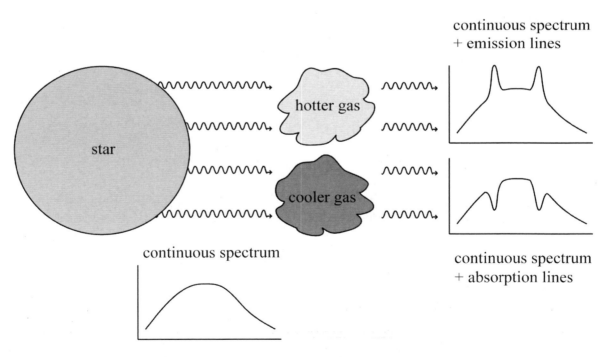

Fig. 10.2-1. Formation of Absorption and Emission Lines.

ionic or molecular species in the upper atmosphere of the star is cooler, more photons are absorbed than are reemitted and an <u>absorption line</u> is produced. If this gas is hotter, more photons are re-emitted than are absorbed and an <u>emission line</u> is produced. Absorption lines constitute the usual case; emission lines require special conditions. Both absorption and emission lines are sensitive to temperature and their strength provides an indication of the temperature of the star.

10.3 The Doppler Effect

If there is relative motion between a source of radiation such as a star and an observer, the wavelength will be shifted toward longer values (redshift) or shorter values (blueshift), according as the source and observer are receding from or approaching each other, respectively. For velocities small compared to the speed of light, the ratio of the shifted wavelength λ' to the unshifted wavelength λ is given approximately by

$$\lambda'/\lambda = 1 + v/c \qquad \text{(Eq. 10.3-1)}$$

where a positive v means recession and a negative v means approach. Taking the logarithm of both sides, we obtain

$$\log (\lambda'/\lambda) = \log (1 + v/c)$$
$$\log \lambda' - \log \lambda = \ln (1 + v/c) / \ln 10$$
$$\Delta\log\lambda = (v/c) / \ln 10$$

where we have used the approximation $\ln (1 + x) = x$ for $x \ll 1$. Finally, we obtain

$$v = \ln 10 \times c \times \Delta\log\lambda \qquad \text{(Eq. 10.3-2)}$$

If the velocity of approach or recession of the portions of gas creating various spectral features is the same, the shifts on a logarithmic wavelength scale will be the same.

10.4 Photometry

Photometry is the measurement of stellar-brightness through a series of filters placed in the observing instrument, which pass a range of wavelengths around an effective wavelength λ_{eff}. The range of wavelengths (full width at half maximum of the transmission curve) is usually 25–30% of λ_{eff}. If flux measurements from a series of filters covering the ultraviolet, visible and infrared portions of the spectrum are put together, a crude spectrum of a star may be obtained. This <u>photometric spectrum</u> does not show the detail of narrow spectral lines, but gives a rough picture of the overall spectrum of the star which may be compared quantitatively with the ideal blackbody spectrum. The common filters with their effective wavelengths are listed in Table 10.4-1.

Table 10.4-1. Common Filters Used for Stellar Photometry.

Ultraviolet (UV)			Visible (VIS)			Infrared (IR)		
Filter	λ_{eff}	$\log \lambda_{eff}$	Filter	λ_{eff}	$\log \lambda_{eff}$	Filter	λ_{eff}	$\log \lambda_{eff}$
M110*	110	2.041	B	440	2.643	I	900	2.954
M133	133	2.124	V	550	2.740	J	1250	3.097
M155	155	2.190	R	700	2.845	H	1650	3.217
M191	191	2.281				K	2170	3.336
M246	246	2.391				L	3540	3.549
M298	298	2.474				M	4800	3.681
U	350	2.544						

*Software filter added to include flux at far end of ultraviolet spectrum.

Taken together, these filters span a factor of 50 in wavelength (a difference of 1.7 in $\log \lambda$) and include nearly all the flux emitted by normal stars.

10.5 Spectroscopy

Detailed stellar spectra are obtained by spreading out the light from a telescope, usually by means of a diffraction grating, and measuring the stellar flux in narrow wavelength channels. The width of these channels, known as the <u>resolution</u> of the spectrum, is sometimes as small as thousandths of a nanometer. Long observing times of many minutes or even hours are necessary to insure a signal-to-noise ratio in each channel which is high enough to obtain high quality spectra. Detailed spectra are usually limited in wavelength range to a factor of two or less, so that compiling a full detailed spectrum of a star requires piecing together measurements from several different instruments. Some of these spectra are shown in Chapter 12.

Table 10.5-1 summarizes the main properties of stars and other celestial objects which can be determined from spectra.

Table 10.5-1. Properties of Stars and Other Celestial Objects Revealed by Spectra.

Property	Spectral Feature Used to Determine Property
Temperature	(1) Location of peak of spectrum
	(2) Relative strengths of spectral lines
Composition	Strengths of spectral lines
Velocity	
(1) Overall	Doppler shift of all spectral lines
(2) Rotational	Doppler broadening of narrow spectral lines
(3) Stellar wind	Doppler blueshift of certain spectral lines

Questions

10-1. Give the equation for Wien's Law and explain what it means.

10-2. Give the equation for the Stefan–Boltzmann law and explain what it means.

10-3. What are the four principal advantages to using the logarithmic wavelength scale?

10-4. What is the difference between color temperature and effective temperature?

10-5. What are the two most important series of spectral lines in hydrogen? Which series has a larger spread on a logarithmic wavelength scale?

10-6. How are absorption and emission lines produced in stars?

10-7. What is the Doppler affect?

10-8. What is the difference between photometry and spectroscopy?

10-9. What does a photometric spectrum show and what does it not show?

10-10. What properties of stars and other celestial objects can be determined from their spectra?

Problems

10-1. A spectrum displayed on a logarithmic wavelength scale has the peak located at log λ_{max} = 2.8. Use Eq.10.1-4 to determine the blackbody temperature.

10-2. The spectrum of a distant object, displaced on a logarithmic wavelength scale, is shifted by an amount $\Delta \log \lambda$ = 0.5 from its rest position. Use Eq.10.3-2 to determine its recession velocity v.

Chapter 11: The Sun

The Sun is a quarter million times closer to us than any other star. We can discern small details on its surface, whereas we struggle with the most sophisticated equipment to detect even the crudest features on the surfaces of other stars. Owing to the Sun's proximity and familiarity, it is fitting that a study of other stars begins with the Sun as a basis for comparison. There are thousands of stars in our galactic neighborhood and millions in our entire galaxy which have a spectral class either identical or similar to the Sun's, and have a similar spectrum. Most of these also appear to have an activity cycle like the Sun's, involving magnetism and spots. However, finding another star which matches the Sun in all respects, in mass, size, surface temperature, chemical composition and age, has not proved to be an easy task. Although solar-type stars share many common features, it appears that they have distinct characteristics as well.

11.1 The Solar Spectrum

Figure 11.1-1 shows a low-resolution spectrum of the Sun.

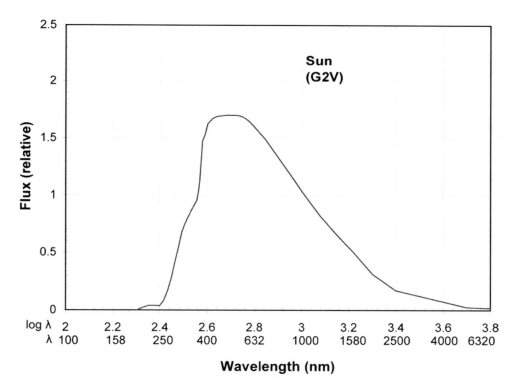

Fig. 11.1-1. The Solar Spectrum.

The spectrum is normalized so that the total area under the curve (using logarithmic wavelength increments times normalized flux increments) is unity. Take the transparency of the blackbody spectrum provided and place it over the solar spectrum. With the wavelength axes coinciding, slide the transparency back and forth until you obtain a best fit by eye between the two curves. You will notice that even at the best-fit position there is a mismatch. The solar spectrum has a steeper slope than the blackbody spectrum at shorter wavelengths and a shallower slope at longer wavelengths. The short-wavelength steepening results from the occurrence of a greater number of absorption lines and edges at shorter wavelengths, i.e. higher photon energies, which rob from the continuous spectrum and hasten its descent.

Example Problem 11.1-1: Read the value of $\log \lambda$ on the solar spectrum which corresponds to the zero point of $\log \lambda$ on the blackbody spectrum. This value is $\log \lambda_{max}$. Use Eq. 10.1-4 to calculate $\log T$ and then T by taking $10^{\log T}$.

You should judge the fit for yourself, but let us say, for example, that you obtain $\log \lambda_{max} = 2.76$. Then,

$$\log T = 6.565 - 2.76 = 3.805$$
$$T = 6380 \text{ K}$$

This temperature represents a color temperature of the solar spectrum.

A quantitative fit between a stellar spectrum and the blackbody spectrum may be obtained by minimizing the <u>difference area</u> between the two curves, where the area is always counted as positive, no matter which curve has the lesser or greater value. For normalized curves like those used here, the difference area represents a fractional mismatch between the two spectra which is zero for a perfect match. A procedure for fitting a blackbody spectrum to a stellar photometric spectrum is given in Appendix A4. When this fitting procedure is executed for the sixteen flux values of the solar spectrum depicted in Fig. 11.1-1, a best fit is achieved with a temperature of 6220 K and a difference area of 0.0909. Therefore, we can say that the solar spectrum deviates from a blackbody spectrum by about 9%.

11.2 Activity Cycles of the Sun and Solar-Type Stars

As stars go, the Sun is a relatively quiet and stable star, which is fortunate for us because the Sun's relative peacefulness has provided an environment in which life on Earth has evolved and flourished. The Sun's departure from quietude occurs mainly in association with an activity cycle, which has three main characteristics:
1. The amount of activity of various forms, such as number of sunspots, solar flares and coronal mass injections increases to a maximum and then decreases to a minimum with a period of 11 +/− 0.5 years.
2. The magnetic fields associated with the solar activity reverse in successive cycles so the full magnetic cycle has a period of 22 years.

3. There are decades-long non-cyclic periods such as the Maunder minimum (1630–1710) when the activity is essentially turned off.

The physical mechanism underlying this activity cycle involves the combination of differential rotation (with both latitude and depth), convection, and meridional (north-south) flow. Reliable information on convection and meridional flow in the solar interior is now available from <u>helioseismology</u>, which analyzes the acoustic oscillations on the solar surface. Quantitative information about the solar interior has led to the development of a predictive model, called the <u>predictive flux-transport dynamo model</u>, which has had good success in predicting the strength and timing of the upcoming solar cycle.

Astronomers have long wondered whether other solar-type stars have similar activity cycles. Because stars are essentially points of light, we cannot observe "starspots" or other features directly, but must rely on detecting variations in the overall brightness level. If we look at the total brightness in all visible wavelengths, the change in the star's brightness between minimum and maximum activity would be only about 0.1%, an amount which cannot be reliably detected through the Earth's atmosphere.

The solution is to look at only a narrow portion of the spectrum where the contrast is greater, and the best region to use is the centers of the H and K lines of ionized calcium at 396.847 nm and 393.366 nm, respectively. Although these lines are mainly <u>absorption</u> lines with an effective width of 1.5 nm and a central brightness which is only 15% of the continuous background, the central 0.1 nm of each line shows an <u>emission</u> peak, due to gas in the chromosphere which is heated by the active regions. If the stellar surface contains many active regions, the central brightness of the line may exceed its quiet level by 30% or more.

Long-term monitoring of the centers of the calcium H and K lines of many solar-type stars since the mid 1980s has shown brightness changes on three time scales:
1. As a star rotates, more active and less active areas of surface emit the light we receive, causing <u>rotational modulation</u> of the brightness, from which the rotation period of the star may be determined. Many of these stars are slow rotators like our Sun and their rotation periods could not be determined by Doppler broadening.
2. As the activity cycle waxes and wanes the average brightness level increases and decreases, and the cycle period may be determined. For most of the stars observed, the periods are in the range of five to fifteen years.
3. For at least one star, namely 54 Piscium, there has been a steady decrease in average activity since 1966, possibly indicating the onset of a Maunder-type minimum.

About one-quarter of the stars observed show a constant low activity and may be in a Maunder-type minimum. In the coming decades, some of these stars may return to a more active state. Long-term monitoring of many solar-type stars is continuing in the northern hemisphere and is now being extended to the southern hemisphere, to include stars such as Alpha Centauri A and B, and Beta Hydri, which are bright enough so that their interior properties can be measured by <u>asteroseismology</u>, the stellar equivalent of helioseismology. Data on these stars will provide new tests on predictive models of cycle activity.

11.3 The Search for Solar Twins

One endeavor which is part of the study of the Sun and solar-type stars is the search for solar twins, i.e. stars whose characteristics are very similar to those of the Sun. Since the Sun has provided a stable environment for life to evolve and flourish on one of its planets, stars very much like the Sun should also provide such stable environments and would therefore be good candidates in the search for extraterrestrial earth-like planets and life. The characteristics of the Sun and the two stars most similar to it found so far are listed in Table 11.4-1.

Table 11.4-1. Characteristics of the Sun and Two Solar Twins.

Star	Sun	18 Scorpii	HD98618 (in UMa)
Spectral Class	G2V	G2V	G5V
Mass (M_o)	1.0	1.01×0.03	1.02×0.03
Luminosity (M_o)	1.0	1.05	1.06×0.05
Temperature (K)	5777	5789	$5843 \times 30K$
Rotation Period (d)	25.4	23.0	
Age (10^9 years)	4.6	4.2	4.3×0.09
Abundances relative to Sun:			
Oxygen	1.0		1.0
Iron	1.0	1.05–1.12	1.11
Co, Cr, Mn, Ni, Ti	1.0		1.0
Lithium	1.0	3.0	3.0
Distance (ly)		45.7	126.0

The main difference between these stars and the Sun is their much greater abundance of lithium, i.e. three times as much. However, it is not expected that this greater portion of lithium would significantly alter the behavior of the star. Although the present data is rather meager, it appears that the chances of any given star in our neighborhood being as similar to the Sun as these two stars are is perhaps one in several thousand. Overall, it is estimated that our galaxy contains perhaps 200 billion stars, so that there may be tens of millions of stars very similar to the Sun.

Questions

11-1. What is the temperature of the blackbody spectrum which best fits the solar spectrum? What is the percent difference between the two?

11-2. What are the main characteristics of the solar cycle?

11-3. What is the underlying physical mechanism of the solar cycle?

11-4. How do astronomers observe solar-type cycles on other stars when they cannot see an image of the star?

11-5. What are the two time scales of the changes in brightness in the calcium H and K lines for the solar-type stars?

11-6. What is the evidence that other stars may be going through a Maunder-type minimum period like the Sun did?

11-7. Why is it important to include the southern hemisphere stars Alpha Centauri A and B and Beta Hydri in studies of solar-type stars?

11-8. Explain the significance of the search for solar twins.

11-9. Approximately what fraction of stars are as similar to the Sun as the most similar ones found so far?

Chapter 12: Stars

In our region of the Milky Way galaxy, stars are far enough apart that from any single star, all the others would appear as mere points of light. It was not until the mid-nineteenth century, nearly two centuries after first getting a good value for an Earth–Sun distance, that astronomers first obtained a good value for an Earth–star distance and realized that interstellar distances are hundreds of thousands to millions of times larger. Nearly all that we know about stars comes from measuring and analyzing their spectra, and this understanding has greatly improved in recent decades by extension of measurements into the ultraviolet and infrared regions of the spectrum, where the very hot and very cool stars, respectively, emit most of their power. Although every star other than the Sun appears to the naked eye as a solitary point of light, such solitude is a deception. We now know that the great majority of stars are accompanied in their journey through space, either by one or more other stars or by planetary companions. It appears that truly solitary stars are rare objects.

12.1 The Stellar Magnitude Scale

The stellar magnitude scale is a logarithmic brightness scale such that a change of one magnitude represents a change of brightness by a factor of the fifth root of 100 (~2.5), and a change of five magnitudes represents a change of brightness by a factor of exactly 100. The magnitude in any filter channel, e.g. the visible (V) filter channel, is defined as

$$m_v = -2.5 \log (f_\lambda / f_{\lambda o}) \qquad \text{(Eq. 12.1-1)}$$

where f_λ is the flux on a linear wavelength scale at the effective wavelength of the filter, in this case 550 nm. The multiplier factor 2.5 is chosen because, on the resulting magnitude scale, the range from the brightest stars (other than Sirius and Canopus) to the faintest naked-eye stars spans about 6 magnitudes, in agreement with the six classes of stellar brightness traditionally depicted in star atlases. The minus sign is chosen so that the fainter stars, which comprise the great majority, have a positive magnitude.

Other magnitudes, such as m_B, m_R, m_I, etc, are defined in a similar manner and the letter m is often omitted so that m_v is written as V and m_B as B, etc. The constant $f_{\lambda o}$ in Eq. 12.1-1 is different for each filter, and is traditionally defined as the flux of the bright star α Lyr (Vega) so that it has zero magnitude in all filter channels. While this definition is convenient to the astronomer who wishes to use Vega as a comparison star, it may be somewhat confusing to the student, because the same magnitude in different filter channels does not represent the same flux level, on either a linear or logarithmic wavelength scale.

In this book, we have converted stellar fluxes f_λ and magnitudes m_V, m_B, etc. to fluxes f in units of $W/m^2/dec$ or some multiple of watts, and display the spectra uniformly on logarithmic wavelength scales. In this way, the true brightness of stars and other objects at any wavelength may be more easily compared.

We define a <u>combined magnitude</u> m, whose zero point represents a round number in metric units:

$$m = -2.5 \log (F/F_o) \qquad \text{(Eq. 12.1-2)}$$

where F is the total or integrated flux over the entire spectrum and $F_o = 25 \times 10^{-9}$ W/m². For stars in the middle range of the spectral sequence, e.g. late A, F, and early G stars, the combined magnitude has a value close to that of the visual magnitude. For O and B stars, as well as K and M stars, it is considerably more negative, because these stars emit most of their energy in the ultraviolet and infrared regions, respectively. The combined magnitude m can be used in any region of the electromagnetic spectrum so that, for example, radio or X-ray sources may be easily compared in brightness with stars.

<u>Example Problem 12.1-1</u>: A star has a total flux of 100 nW/m². What is its combined magnitude?

$$m = -2.5 \log (100 \times 10^{-9} \text{ W/m}^2 / 25 \times 10^{-9} \text{ W/m}^2)$$
$$m = -2.5 \log (4)$$
$$\boxed{m = -1.51}$$

Appendix A8 is a list by spectral class of all stars having a combined magnitude less than 2.5, i.e. a total flux greater than 2.5×10^{-9} W/m². There are approximately 300 stars in all. The fluxes in the ultraviolet (91.15–400 nm), visible (400–800 nm) and infrared (800–5000 nm) regions are calculated by averaging the fluxes in the filter channels of each region. (For this calculation, the visible region is taken as 400–800 nm instead of 400–700 nm so as to include completely the contribution of the R filter). Note the much greater preponderance of hot blue stars (O and B) and cool red stars (M) as compared with the usual lists of bright stars because of the large contributions to their total flux from their ultraviolet and infrared regions, respectively. Note also that α Cru (Acrux) has displaced α CMa (Sirius) as the brightest star.

12.2 Parallax, Distance and Absolute Magnitude

In measuring distances to the nearby stars, we follow the same procedure discussed in Section 2.8, expect that the baseline D is now 2 AU. Equation 2.8-1 can be written as

$$d = 1/p \qquad \text{(Eq. 12.2-1)}$$

where the distance d is in AU and the parallax p in radians. For interstellar measurements, it is more convenient to measure the parallax p in arc seconds (1/206265 radian) and d in a unit called a parsec (206265 AU), abbreviated as pc. Because d and p are reciprocally related, Eq. 12.2-1 is still valid in these units.

Measuring stellar parallax requires the combination of the telescope and photography and the first such parallax measured was that of the binary star 61 Cyg by Friedrich Bessel in Prussia in 1838. Due to the blurring of the stellar images by atmospheric turbulence, reliable values for parallax by ground-based observations could be obtained only for nearby stars. With the launching of the Hipparcos (<u>hi</u>gh <u>p</u>recision <u>par</u>allax <u>c</u>ollecting <u>s</u>atellite) by the European Space Agency in 1989, accuracy improved by more than an order of magnitude. The uncertainty in the parallax values is now a few parts per thousand for nearby stars and 10–20% for stars at a distance of 200 pc.

Appendix A7 is a list of the twenty nearest star systems including the Sun. It provides a representative sample of the stellar population in our galactic neighborhood. Of the twenty star systems, twelve (60%) have a low-mass, low-luminosity class MV red dwarf as the primary star. An additional six red dwarfs are present as secondary or tertiary components, so that, overall, red dwarfs comprise 18 out of 30 (60%) of the stars listed. The preponderance of red dwarfs reflects the fact that when stars are born in clusters out of gas clouds, low-mass stars are most probable, with higher mass ones less likely the higher the mass.

When mankind eventually sets up bases in the outer solar system, the measurement of stellar parallax will again improve dramatically. From Neptune, with a baseline 30 times larger than from Earth, the accuracy would improve by a factor of thirty and allow us to obtain an accurate three-dimensional star map of a good portion of the Milky Way Galaxy.

The magnitudes discussed so far are <u>apparent magnitudes</u>, i.e. those related to the brightness or flux observed from Earth. We also define an <u>absolute magnitude</u>, related to the brightness or flux observed at a standard distance of 10 parsecs. Analogous to Eq. 12.11, we define the absolute visual magnitude M_V as

$$M_V = 2.5 \log (f_{\lambda 10}/f_{\lambda o}) \qquad \text{(Eq. 12.2-1)}$$

where $f_{\lambda 10}$ is the flux observed at 10 parsecs. $f_{\lambda 10}$ is related to f_λ, the actual flux observed when the star is at its true distance d (in parsecs) by the inverse square law:

$$f_\lambda/f_{\lambda 10} = (10/d)^2 \qquad \text{(Eq. 12.2-2)}$$

Combining Eqs. 12.1-1, 12.2-1, and 12.2-2 to eliminate the fluxes, we derive a very important equation relating distance, apparent magnitude and absolute magnitude

$$\log d = (m_V - M_V)/5 + 1 \qquad \text{(Eq. 12.2-3)}$$

where the distance d is in parsecs. This equation forms the basis for all distance determinations using "standard candles", e.g. RR Lyrae variables, Cepheid variables, and Type Ia supernovae, at distances beyond the range of parallax measurements.

Example Problem 12.2-1: The star α Cyg (Deneb) has an apparent visual magnitude m_V of 1.26 and an estimated absolute visual magnitude M_V of -8.7. What is its distance?

$$\log d = (1.26 + 8.7)/5 + 1 = 2.99$$

$$\boxed{d = 977 \text{ pc}}$$

12.3 Photometric Spectra of Stars

Photometric spectra of some bright stars of different spectral types are shown in Figs. 12.3-1 to 12.3-3. Figure 12.3-1a shows the spectrum of α Ori (Betelgeuse) in absolute flux units in which the total flux from 91.15 nm to 5000 nm is 83.93 nW/m^2. Figure 12.3-1b shows the normalized spectrum where the total area (including an estimated 3% contribution from wavelengths longer than 5000 nm) is unity. If you place the blackbody spectrum transparency over the normalized spectrum and adjust it to achieve the best fit by eye, you will see that a fairly good fit can be achieved. A quantitative fit using the procedure of minimizing the difference area (see Appendix A4) yields a best fit at a temperature of 3150 K and a difference area of .0791, so that the fit is somewhat better than for the Sun.

Figure 12.3-2 shows normalized spectra for two A stars, namely α Aql (Altair) and α Lyr (Vega), which are hotter than the Sun. Here, significant departures from the blackbody spectrum can be seen. For Altair, the peak of the spectrum is noticeably narrower and higher than the peak for the blackbody. The same is true for Vega, and in addition, a second peak is present in the far ultraviolet. What is happening is that at wavelengths shorter than the Balmer limit at 364.6 nm cooler hydrogen gas in the upper atmosphere of the stars is robbing significantly from the main peak on the ultraviolet side and narrowing it. The abrupt drop in flux at the Balmer limit is called the Balmer jump. The deviations from the blackbody spectrum, as measured by the difference area are significantly higher, being .1962 and .3149, respectively.

Figure 12.3-3 shows normalized spectra for two B stars, namely δ Per and η UMa (Alkaid), which are still hotter. The spectrum shifts toward the ultraviolet and the far ultraviolet peak becomes the dominant one. At hotter temperatures, more hydrogen in the stellar atmosphere becomes ionized, and the Balmer jump effect weakens. Deviations from the blackbody spectrum decrease somewhat, the difference areas for these two stars being .2651 and .2371, respectively.

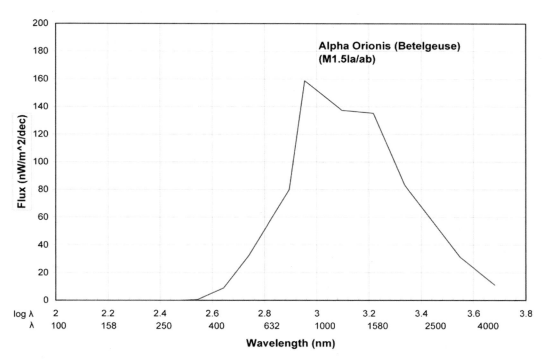

Fig. 12.3-1a. Photometric Spectrum of Alpha Orionis (Betelgeuse) (absolute).

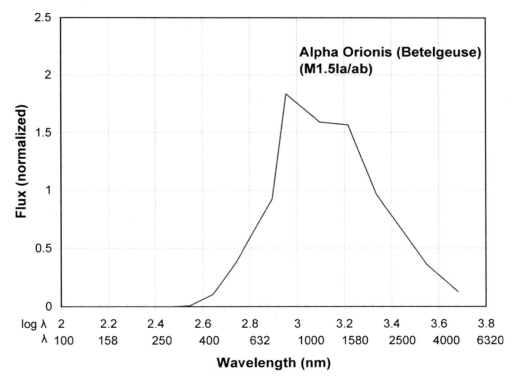

Fig. 12.3-1b. Photometric Spectrum of Alpha Orionis (Betelgeuse) (normalized).

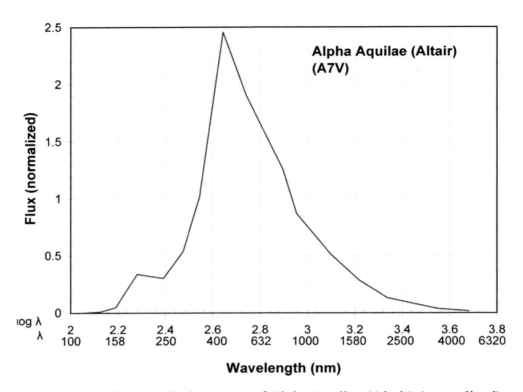

Fig. 12.3-2a. Photometric Spectrum of Alpha Aquilae (Altair) (normalized).

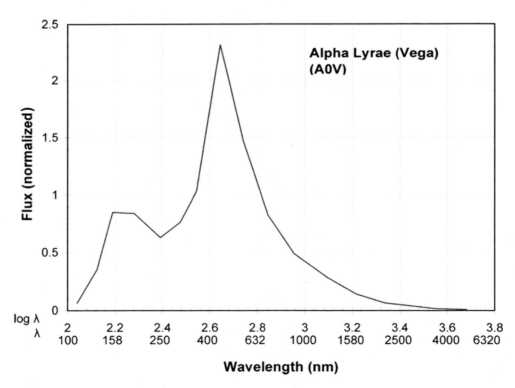

Fig. 12.3-2b. Photometric Spectrum of Alpha Lyrae (Vega) (normalized).

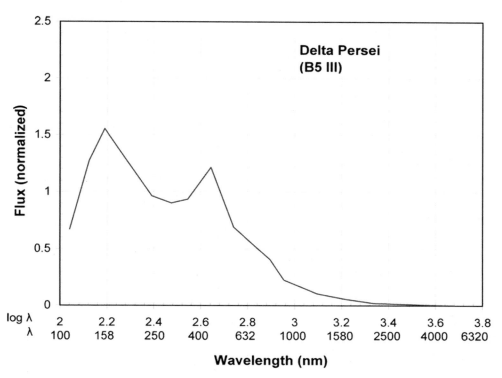

Fig. 12.3-3a. Photometric Spectrum of Delta Persei (normalized).

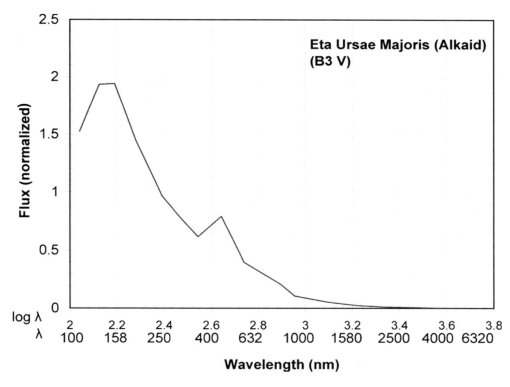

Fig. 12.3-3b. Photometric Spectrum of Eta Ursae Majoris (Alkaid) (normalized).

The fractional deviation from a blackbody spectrum versus peak wavelength for the preceding and some other stars spanning the entire range of stellar spectra is shown in Fig. 12.3-4. The closest fit to a blackbody spectrum occurs between the Sun's spectral class and the early M stars. The late M stars are so cool that their atmospheres contain many molecules, such as TiO_2, which have absorption bands which rob from the spectrum. The very hottest O stars have significant flux above the Lyman limit at 91.15 nm, which is emitted by the star but absorbed in interstellar space so that the observed spectrum departs significantly from a blackbody spectrum. Over the range shown, the greatest departure occurs for stars like Vega, of spectral class A0, where the Balmer jump effect is at its maximum.

The procedure for fitting a stellar spectrum with a blackbody spectrum, both by eye and using a computer program is illustrated in Laboratory Exercise LE3.

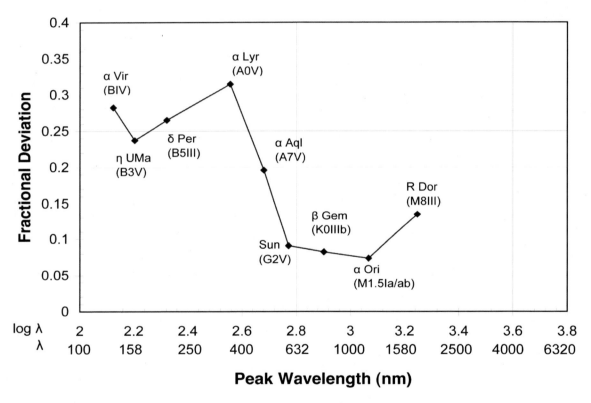

Fig. 12.3-4. Fractional Deviation from Blackbody Spectrum versus Peak Wavelength.

12.4 Ultraviolet Spectra of Stars

For many decades ground-based stellar astronomy had to make do with observations in the visible portion of the spectrum recorded on photographic plates, and the science of stellar spectral classification was developed based on these observations. However, for the very hot O and B stars, features in the visible part of the spectrum are rather weak. The strong and interesting features occur in the ultraviolet portion of the spectrum. Beginning in the 1960s, ultraviolet satellites have opened this region to observation and the results have given us a much better picture of the early-type stars. Some of these results are presented in this section.

Figure 12.4-1a shows the entire ultraviolet spectrum of α Vir (Spica) from the Lyman limit at 91.15 nm up to nearly the Balmer limit at 364.6 nm, a range of a factor of four in wavelength, which is 2.5 times the range of the visible region. The spectrum is pieced together from four sources: EURD (Espectrografo Ultravioleta extreme para la Radiacion Difusa) for 91.15 nm to 100 nm, Copernicus for 100 nm to 120 nm, IUE (International Ultraviolet Explorer) for 120 to 195 nm, and OAO2 (Orbiting Astophysical Observatory 2) for 195 nm to 360 nm. The spectral resolution for the EURD and OAO2 data is lower than for the Copernicus and IUE data, so the spectrum looks smoother in these regions. The peak of the spectrum occurs at a wavelength of ~110 nm.

Figure 12.4-1b shows the spectrum for the star σ Sco taken from the same data sources, except that the portion below 100 nm, which has not been actually measured, is estimated from the results for α Vir. These two stars have nearly identical spectral types, but their spectra are noticeably different. For σ Sco, the peak of the spectrum is at ~130 nm and there is very little flux below 100 nm. A broad absorption feature with a maximum at ~220 nm is also evident. These features are due to interstellar extinction in the form of both absorption and scattering by gas and dust, which is most prominent in the ultraviolet. The extinction is much stronger for σ Sco, which lies in the plane of the galactic disk than for α Vir, which lies out of it. Because of interstellar extinction, the proportion of O and B stars in the stars of a given magnitude range, e.g. 4 to 5, gets smaller as the magnitude numbers increase.

Absorption lines at the Lyman series wavelengths (121.53 nm, 102.54 nm, 97.23 nm, and 94.95 nm) mostly due to absorption by interstellar hydrogen are evident, particularly in the α Vir spectrum. Most of the other important spectral lines are due to ionized forms of three elements: carbon, nitrogen, and silicon. As is the case with lines of hydrogen and other elements in the visible region, these lines have a maximum intensity at some temperature and diminish on either side. They are also usually sensitive to stellar luminosity.

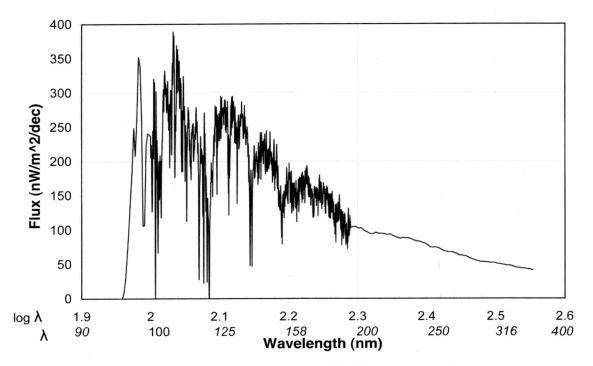

Fig. 12.4-1a. Ultraviolet Spectrum of Alpha Virginis (Spica).

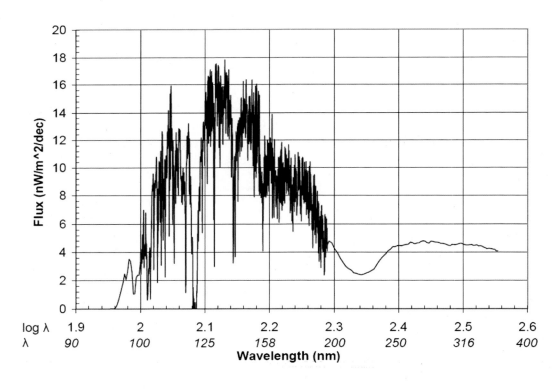

Fig. 12.4-1b. Ultraviolet Spectrum of Sigma Scorpii.

The most important ultraviolet spectral lines are listed in Table 12.4-1. The Roman numeral following the chemical symbol is one more than the number of electrons missing in the ion, e.g. C IV is carbon with three electrons missing.

Table 12.4-1. Important Ultraviolet Spectral Lines.

Ion	Wavelengths	Max. @	Sensitivity to	
			Spectral Type	Luminosity
C III	117.5,117.6	B1	Yes	No
N V	123.9, 124.3	mid O	Yes	Yes
Si II	126.0, 126.5	late B	Yes	
Si III	129.5, 130.3	early B	Yes	Slight
Si II	130.4, 130.9	late B	Yes	
Si IV	139.4,140.3	B 1	Yes	Very
C IV	154.8,155.1	mid O	Yes	Very
N IV	171.9	mid O	No	Yes

One distinguishing feature of the spectra of the hottest and most luminous stars is the presence of combination <u>absorption–emission</u> lines, often called P Cyg lines after the prototype star. The mechanism of formation is depicted in Fig. 12.4-2.

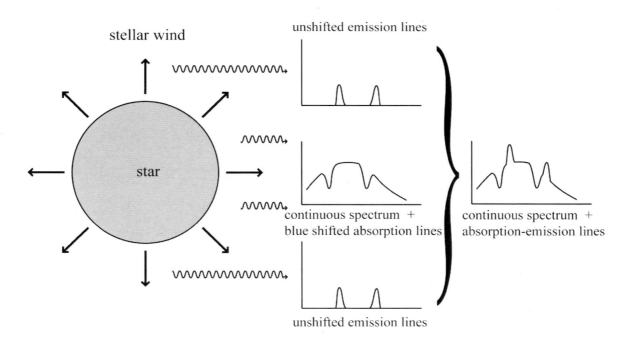

Fig. 12.4-2. Formation of Absorption-Emission Lines.

These lines are striking features in the spectra. On the short-wavelength side, there is an absorption line, sometimes reaching total absorption, and on the long-wavelength side, there is an emission line whose peak is sometimes more than twice the continuum background level. They are produced by very strong stellar winds driven by the intense radiation pressure in these very hot stars. Flux coming from the stellar wind in front of the star is emitted by gas cooler than the star, is seen in <u>absorption</u>, and is <u>blueshifted</u> because the gas is approaching us. Flux emitted by the gas at the side of the star has nothing in back of it, is seen in <u>emission</u> and is <u>unshifted</u> because the gas is moving transversely. From the degree of blueshift, the velocity of the stellar winds can be estimated.

Figures 12.4-3a–12.4-3d show a series of spectra of stars having identical or nearly identical luminosity classes (Ia, Iab) and spectral classes of decreasing temperature. The Si IV absorption–emission line group at ~140nm log λ ~ 2.142 − 2.147(two absorption lines, one emission line) increases to a maximum at 08 (9 Sge) and then decreases until the emission is nearly gone at B3 (o^2 CMa). The amount of blueshift steadily decreases as the cooler temperatures produce a lower velocity of the stellar wind.

The analysis of absorption–emission lines to understand stellar winds is a classic example of how astronomers construct a plausible physical model to explain a stellar spectral feature, although the star appears to us as a mere point of light.

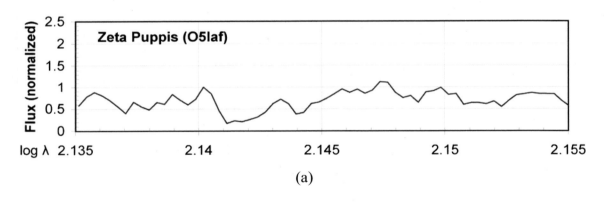

(a)

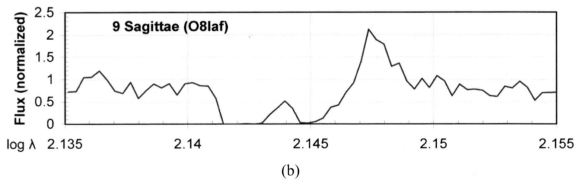

(b)

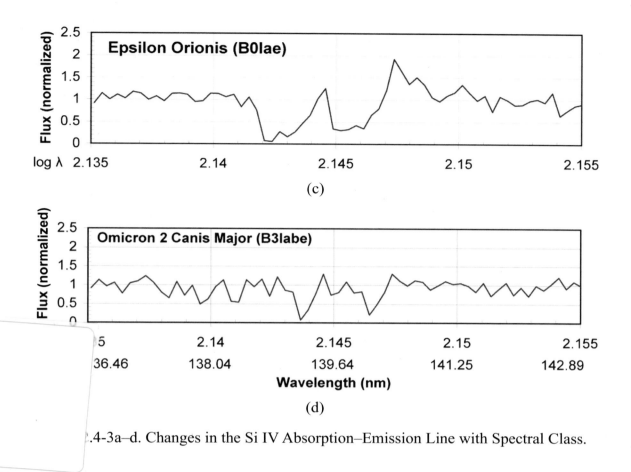

Fig. 12.4-3a–d. Changes in the Si IV Absorption–Emission Line with Spectral Class.

12.5 Visual Binary Stars

In the list of the twenty nearest star systems in Appendix A7, eight of the systems are multiple. Six are double, one (Luyten 789-6) is triple and one (α Cen) is suspected to be triple. Even from this small sample, it is clear that multiple star systems are common. Luyten 789-6 contains a spectroscopic binary with a period of 3.8 days. If α Cen C (Proxima) does orbit the AB binary, its period would be at least hundreds of thousands of years. The other eight orbital periods range from 2.25 years to 2600 years, with a median value of 65 years. This range is comparable to the range of periods of the planets, asteroids, and dwarf planets in our solar system.

These <u>visual binary stars</u>, which have been observed to complete one or more orbits or at least enough of an orbit to determine it, all travel in elliptical orbits about one another. Because the plane of the orbit is randomly distributed with respect to the line of sight, the orbit we see, although it is elliptical in shape, is not the true orbit but a projected view. However, the true orbit may be determined by a straightforward procedure.

Figure 12.5-1 shows a projected view of the orbit which was shown viewed pole-on in Fig. 2.5-1. In order to create the projected view, we must first <u>rotate</u> the major axes with respect to the x (horizontal) axis by angle φ (either positive or negative) and then <u>tilt</u> the x–y plane about the x-axis by angle θ. During this process the apparent lengths of the major and minor axes may decrease and the angle between them, δ, depart from 90°. However, the ratio $CF_1/CA = CF_1/CP$, which is the eccentricity e of the orbit, remains invariant. In addition, the lines tangent to the orbit at the minor-axis endpoints B_1 and B_2 remain parallel to the major axis.

Of course, we do not observe the major axis; we observe the projected elliptical orbit and the focus F_1, where the companion star is located. The major axis may be constructed with the aid of a <u>parallel ruler device</u> by pivoting the center line about F_1 and adjusting the parallel rulers until they are equidistant from the center line and tangent to the orbit. Once the major axis is drawn and the points A, P, and C determined, the eccentricity e of the <u>true orbit</u> may be calculated as $CF_1/CA = CF_1/CP$. The axial ratio b/a of the <u>true orbit</u> is then equal to $\sqrt{(1 - e^2)}$.

The minor axis may be constructed as the line which passes through the center C and intersects the orbit at the tangent points B_1 and B_2. The semimajor axis a and semiminor axis b of the projected ellipse may be measured and the axial ratio b′/a′ of the <u>apparent orbit</u> calculated. Both the axial ratio quotient

$$q = (b'/a') / (b/a) \qquad \text{(Eq. 12.5-1)}$$

and the apparent major-to-minor axis angle δ′ are functions of the rotation angle φ and the tilt angle θ as given in Appendix A2. These equations cannot be inverted to obtain expressions for φ and θ, but an interpolation graph may be constructed to determine φ and θ numerically and obtain the true semimajor axis a.

The procedure for obtaining the true orbits of visual binary stars is illustrated in Laboratory Exercise LE2.

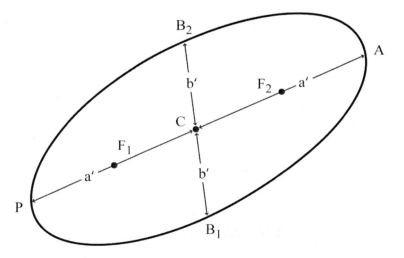

Fig. 12.5-1. Elliptical Orbit Rotated and Tilted.

Questions

12-1.   Why is the factor 2.5 used in the stellar magnitude scale?

12-2.   What is the combined magnitude? How does it compare with the visual magnitude for stars of different spectral classes?

12-3.   Why are there many more O, B, and M stars in a list of bright stars in combined magnitude as compared with a list based on visual magnitude?

12-4.   What is a parsec?

12-5.   Why was not stellar parallax measured until 1838?

12-6.   What is the most common type of star in our stellar neighborhood?

12-7.   What is the absolute magnitude of a star?

12-8.   What is the Balmer jump?

12-9.   For which spectral types of stars is the fit to the spectrum by a blackbody spectrum the best? The worst?

12-10. What features in the ultraviolet spectra of stars are introduced by interstellar extinction?

12-11. What three chemical elements produce most of the absorption lines seen in stellar ultraviolet spectra?

12-12. Explain how combination absorption–emission lines are produced.

12-13. How can the major axis of the true orbit of a binary star be constructed from the observed projected ellipse?

12-14. What two equations are necessary to determine the individual masses of a binary star system? (See Steps 12 and 13 in Laboratory Exercise LE2)

Problems

12-1.   A star has a total flux of 25 nW/m². What is its combined magnitude?

12-2.   A star has a parallax of .25 arc seconds. What is its distance in parsecs and in light years?

12-3.   The star β Ori (Rigel) has an apparent visual magnitude m_v of 0.12 and an estimated absolute visual magnitude M_v of -7.0. What is its distance?

Chapter 13: Stellar Evolution

All stars, after being born in a group in an interstellar gas cloud, spend the major portion of their lives on the main sequence fusing hydrogen to helium. However, the length in years of this main-sequence stage varies enormously, ranging from hundreds of billions of years for the low-mass cool red dwarfs to only a few million years for the high-mass hot blue stars. After leaving the main sequence, most stars go through pulsating and, for the highest-mass stars, explosive phases. In these phases, the stars are bright and, in some cases, can be used as standard candles to gauge the vast distances to faraway galaxies. Through stellar evolution, nature has given us a way of determining the size and fate of our universe.

13.1 Mass Determines a Star's Fate

A star's fate is governed mainly by its mass, which determines its core temperature, its luminosity, and the type of nuclear reactions which will occur in the core. Although all the details are not completely understood, it appears that there are basically five ranges of stellar mass each having a distinct type of evolution after the main-sequence stage. These are listed in Table 13.1-1.

Table 13.1-1. Mass Ranges and Post-Main-Sequence Evolution of Stars.

Mass Range (M_O)	Main Sequence Lifetime (10^9 yr)	Post-Main-Sequence Evolution
.08-.4	> 100	No helium burning – cooling
.4-2	100–1.8	Sudden onset of helium burning → carbon burning → planetary nebula → white dwarf
2-8	1.8 – .0055	Gradual onset of helium burning → carbon burning → planetary nebula → white dwarf
8-25	.0055 – .0003	Helium/carbon/silicon burning → iron → Type II supernova → neutron star
>25	<.0003	Helium/carbon/silicon burning → iron → Type II supernova → black hole

For stars of the lowest mass range, i.e. the red dwarfs, the star is fully convective and the helium, instead of accumulating in the core, is mixed throughout the star. It never gets hot enough to initiate helium burning and, when all the hydrogen has been converted to helium, the star will simply cool down gradually. The time scales for this process are so long that no star in this mass range has yet finished its hydrogen burning.

For stars of the second and third mass ranges, helium burning, followed by carbon burning, eventually occurs, but it is initiated in different ways. In stars having a mass similar to the Sun's (.4–2 M_O), gravitational contraction is not strong enough to raise the temperature of the helium core to the burning point. Helium burning commences only when the core becomes degenerate, shrinks and initates a sudden burning known as the <u>helium flash</u>. For stars of the third mass range (2–8 M_o), the gravitational contraction is strong enough to initiate helium burning slowly. These stars move off the main sequence through the instability strip and become Cepheid variables. However, the ultimate fate of all stars with masses between $.4M_o$ and $8M_o$ is the same: they lose their outer layers as planetary nebulae and ultimately become white dwarfs.

For stars of the fourth and fifth mass ranges, gravitational contraction is strong enough to raise the core temperature high enough so that helium burning followed by carbon burning followed by silicon burning to iron occurs. The buildup of an unstable iron core results in a Type II supernova. The difference is that, for stars with 8–25 M_o, only enough material is left in the core (1.4–3 M_o) to form a neutron star, whereas for stars with >25 M_o, the residual core mass exceeds 3 M_o and forms a black hole.

Two other factors have an effect on a star's fate. The first is its metallicity, i.e. its content of elements other than hydrogen and helium. Metallicites of stars range from less than one percent of the Sun's to several times the Sun's. Higher metallicity means greater opacity of the gas to radiation, which can affect the star's evolution, particularly in the pulsating phase.

The second factor is the presence of a close companion in binary or multiple star systems. If the companion evolves faster, it may expand into the red giant phase and transfer material to the given star, altering its evolution. If the given star evolves faster to the white dwarf stage, when the companion star evolves into the red giant phase, it may transfer material to the given star and a nova or Type Ia supernova may result (See Section 13.3).

13.2 Pulsating Stars

Pulsating stars with moderate to large amplitudes (a change in visual magnitude by .35 or more) are giants which occur in the later stages of stellar evolution. The pulsations are driven by a layer of partially ionized helium or another element which absorbs radiation, expands and becomes more transparent, releases radiation, contracts and becomes more opaque, etc., in a cyclical fashion. Pressure and temperature conditions are favorable for sustaining the oscillations, which would be damped out in a denser main-sequence star. There are three main types.

RR Lyrae Variables (yellow giants)

RR Lyrae variables, named after the prototype star RR Lyrae, are old stars, originally a bit less massive than the Sun, but now at $\sim.5$ M_o, having lost significant mass in the red giant phase. They are prevalent in stellar populations with low metal content (Population II) such as the older globular clusters and the galactic halo. Most pulsate in the fundamental mode with a typical period of a bit more than .5 day and an amplitude of ~1 magnitude. A smaller number pulsate in an overtone mode with a period of $\sim.3$ day and an amplitude only half as large. Their average absolute magnitude is $\sim.75$, meaning that their luminosity is ~45 L_O. Harlow Shapley used the RR Lyrae variables to measure the distance to about 70 globular clusters and show that the Sun's position was significantly displaced from the center of the Milky Way galaxy. Because of their narrow range of periods and luminosities, the RR Lyrae variables are good standard candles, but their relatively low luminosity limits their use to the globular clusters and satellite galaxies of the Milky Way.

Cepheid Variables (yellow giants)

The Type I Cepheid variables, named after the prototype star Delta Cephei, are young stars with masses in the range 2.5–8 M_o and periods in the range 2–50 days. They were originally main-sequence A3–B4 stars and have evolved off it and are crossing the instability strip in the H–R diagram where pulsations are sustained. Stars of higher mass and luminosity are larger and take longer to pulsate giving rise to a period-luminosity relation which can be expressed as:

$$M_V = M_{V0} + s \log P \qquad \text{(Eq. 13.2-1)}$$

The P–L relation and their much larger luminosities make the Cepheids very useful as distance indicators if the zero point M_{V0} can be determined. The difficulty is that no Cepheid is near enough to have its distance accurately determined by parallax, so that calibration has to employ statistical methods involving the Sun's peculiar velocity through the galaxy, which have a large uncertainty.

Due to systematic errors and a failure to include the proper correction for interstellar extinction, Harlow Shapley underestimated the true brightness of the Type I Cepheids by ~1.5 magnitudes, i.e. a factor of four. However, this discrepancy was not noticed for several decades. There exists a second kind of Cepheid variable known as the Type II Cepheids or W Virginis stars, which also occur in the globular clusters. They are typically fainter than the Type I Cepheids by the same 1.5 magnitudes, i.e. a factor of 4. The observed relative apparent magnitudes of RR Lyrae variables and Type II Cepheid variables (mistakenly thought to be Type I) were in agreement with the relative absolute magnitudes of the RR Lyrae variables and (mistakenly estimated) values for the Type I Cepheids. The two errors cancelled almost exactly!

The discrepancy became apparent in 1948 when the Andromeda galaxy was first observed with the 200-inch Mt. Palomar telescope. Cepheid variables were clearly observed in the galactic disk but RR Lyrae variables, which should have been visible if they were the same brightness relative to the Cepheids as in the Milky Way's globular clusters, were not. It was Walter Baade who recognized that there are two distinct stellar populations and that Cepheid variables within them are also distinct: the Type II Cepheids typically being fainter than the Type I of the same period by 1.5 magnitudes. In one stroke, our value of the distance to the Andromeda galaxy, and to every other galaxy, doubled!

Since that time there have been further refinements of the Cepheid calibration taking into account interstellar reddening effects using parallaxes measured by Hipparcos. Modern values of the absolute magnitude, luminosity, and mass of both types of Cepheids are given in Table 13.2-1.

Table 13.2-1. Absolute Magnitudes, Luminosities, and Masses for Cepheid Variables.

log P	P (d)	Type I Cepheids					Type II Cepheids			
		M_V	log L	L	log M	M	M_V	log L	L	M
0.0	1.0	−1.43	2.31	325	.39	2.45	−0.2	2.02	105	.60
0.2	1.58	−1.99	2.74	540	.44	2.75	−0.6	2.18	150	.59
0.4	2.5	−2.55	2.96	910	.50	3.16	−1.0	2.34	220	.58
0.6	4.0	−3.12	3.19	1540	.56	3.63	−1.3	2.46	290	.57
0.8	6.3	−3.68	3.41	2580	.61	4.07	−1.7	2.62	420	.56
1.0	10.0	−4.24	3.64	4325	.67	4.68	−2.1	2.78	600	.55
1.2	15.8	−4.80	3.86	7240	.72	5.25	−2.5	2.94	870	.54
1.4	25.0	−5.36	4.08	12130	.78	6.03	−2.9	3.10	1260	.53
1.6	40.0	−5.93	4.31	20510	.83	6.76	−3.2	3.22	1660	.52
1.8	63.0	−6.49	4.54	34360	.89	7.76	−3.6	3.38	2400	.51

Note that, for the Type I Cepheids, there also exists a _mass–luminosity_ (M-L) relation of the form

$$\log L = \log L_0 + s \log M \qquad \text{(Eq. 13.2-2)}$$

It is left as a problem (Prob. 13-1) to determine the intercept $\log L_0$ and the slope s. This relation is different from the M–L relation for main-sequence stars where the slope is ~3.5 and the intercept is zero. Cepheids are giant stars so that, for a given mass, the luminosity is much higher than for main-sequence stars.

It is also clear from the table that the Type II Cepheids are a totally different kind of star. Like the RR Lyrae variables, they are old stars of low mass, having previously lost significant mass in a red-giant phase. The _decrease_ in mass with luminosity suggests that they may be ascending the asymptotic giant branch (AGB) of the H–R diagram, losing mass, increasing in luminosity and lengthening in period, until they reach the tip of AGB, move across the instability strip, and become planetary nebulae and white dwarfs.

Example Problem 13.2-1: From Table 13.2-1, find the values of the zero point M_{V0} and the slope s for both the Type I and Type II Cepheids. Are the Type II Cepheids fainter by 1.5 magnitudes for all values of log P? If not, at what value of log P is the difference in M_V closest to 1.5?

Type I = M_{V0} = 1.43   s = (–6.49 – (–1.43))/1.8 = –2.81
Type II = M_{V0} = –0.2   s = (–3.6 – (–0.2))/1.8 = –1.9
No, because the slopes are different
At log P = 0.4, M_V (Type II) – M_V (Type I) = –1.0 – (–2.55) = 1.55

Long-Period Variables (red giants)

Long-period variables include Mira variables, named after the prototype star Mira (Omicron Ceti) and semi-regular variables. These stars originate from main-sequence stars with masses somewhat greater than 1 M_o which are evolving along the asymptotic giant branch (AGB) and will experience significant mass loss before returning to the yellow-giant region. They are best observed in the infrared, where the peak of the spectrum is located, and where the pulsations are more uniform than in the visible region. Miras have fairly regular pulsations in the range 100–400 days. Semi-regular variables usually have shorter periods, smaller amplitudes, and more irregular pulsations than Miras.

Studies of Miras in the Large Magellanic Cloud and both Miras and semi-regular variables in the galactic field stars with accurate distances measured by Hipparcos indicate that these stars may obey two period-luminosity relations of the form

$$M_K = M_{K0} + s \log P \qquad \text{(Eq. 13.2-3)}$$

where M_K is the absolute magnitude in the K band. The steeper line has a slope ~25% steeper than the line for the Type I Cepheids, while the shallower line has a slope only ~60% as large. The occurrence of two period–luminosity relations may be due to a difference in mode, e.g. fundamental versus overtone, or a difference in stellar structure. Despite their fuzzier period–luminosity relation and the longer observing times required to measure their periods, the long-period variables have already proven useful as distance indicators, and their use will undoubtedly increase in the future.

13.3 Novae and Type Ia Supernovae

Periodic variable stars were discovered relatively recently in the whole span of astronomical observation: Miras in 1596 and Cepheids in 1684. In contrast, eruptive variables, i.e. stars which appear out of nowhere, shine brightly for a few weeks or months, and then gradually fade from view, were known since antiquity. Chinese astronomers witnessed a "guest star" in the constellation Taurus in 1054 and Tycho Brahe observed a similar event in Cassiopeia in 1572. Orginally, all such eruptive stellar events were called 'novae' from the Latin phrase 'nova stella' ('new star'). However, when Edwin Hubble discovered Cepheid variables in the Andromeda Galaxy M31 in 1924 and calculated its distance (too small at the time because of the mistake with the Cepheids, but still well outside the Milky Way), it was realized that the eruptive event S Andromedae, observed in 1885, had to be much brighter than the novae in our own galaxy and there was a separate class of eruptive events called 'supernovae'. Novae are typically ~100,000 times brighter than the Sun and ~20 per year occur in the Milky Way. Supernovae are typically several billion times brighter than the Sun and maybe, on average, one occurs in the Milky Way every 50 years. However, most events of either type are invisible to us, hidden by the dust in the plane of the Milky Way's disk.

It is now thought that both novae and Type Ia supernovae occur in binary star systems by accretion of hydrogen gas from a red giant onto a white dwarf. In the case of a nova, the hydrogen builds up in a layer to a thickness of ~100 m on the surface of the white dwarf. When the pressure and temperature at the bottom of the layer reach a critical point, hydrogen fusion is suddenly initiated and the entire layer explodes, sending out a nova shell and increasing the brightness of the binary system (mostly due to the red giant) by many times. Having acted as the catalyst for the explosion, the white dwarf remains intact to repeat the process. Many novae are observed to be recurrent on the scale of decades, although they do not follow a strict periodicity.

In the case of a Type Ia supernova, the accreting hydrogen fuses continually to helium and gradually increases the mass of the white dwarf. When the mass reaches the Chandresekhar limit of 1.38 M_o, the dwarf collapses, carbon fusion is initiated and quickly propagates through the dwarf and the entire star explodes. The expanding cloud of heavier elements contains a large amount of iron, cobalt, and nickel produced in the nuclear reactions, but very little hydrogen. No remnant is left at the center. A typical Type Ia supernova spectrum is shown in Fig. 13.3-1a. There are no hydrogen lines. The principal absorption lines are due to ionized forms of silicon, iron, and cobalt, with the doubly ionized silicon (Si II) line at a rest wavelength of 653.5 nm being prominent.

The positions of the lines are due to two spectral shifts: (1) a blueshift due to the motion of the absorbing gas, seen in front of the exploding star, approaching us at expulsion speeds of up to 10000 km/s, and (2) a redshift due to the expansion of the universe which increases as the supernova's distance increases. Both the shifts must be considered when analyzing the spectrum. Note that the spectrum is very different from a blackbody spectrum.

The intriguing question has always been: Why do some white-dwarf binary systems produce novae and others produce Type Ia supernovae? Although the conditions which distinguish between these two types of events are not yet fully understood, it does appear that the rate of mass transfer of the hydrogen gas to the white dwarf is a key factor. At relatively low mass transfer rates, the hydrogen gas remains relatively cool and fusion is initiated suddenly only when a critical mass is reached. Essentially all the hydrogen is blown off the dwarf so that little mass is added. As long as the companion remains in the red-giant phase, nova explosions may recur, but the dwarf will never reach the Chandresekhar limit. At moderate mass transfer rates, the hydrogen gas heats up, fusion occurs continually and the helium byproduct adds to the white dwarf's mass, in some cases, eventually pushing it to the Chandresekhar limit. Theoretical studies indicate that, at still higher mass transfer rates, a stellar wind develops and much of the transferred hydrogen is blown off into space. However, the mass of the white dwarf still increases, and eventually the red giant's mass decreases significantly and the mass transfer rate drops to a moderate level. A Type Ia supernova may still eventually occur.

Because Type Ia supernovae all result from the same type of exploding star, namely a white dwarf at the Chandresekhar limit, their intrinsic brightnesses are all very similar, and a standard value for the absolute magnitude is taken to be –19.5. Because of their importance as standard candles at great distances, much work has been devoted to their calibration. Although their actual absolute magnitudes may vary in either direction by up to 1.5 magnitude from the standard value, this variation may be largely corrected for by observing the <u>time dependence</u> of their magnitudes, i.e. their <u>light curves</u>. Brighter Type Ia supernovae take longer to rise to the maximum and fall from it than do dimmer ones.

13.4 Type II Supernovae

Type II supernovae result from the explosion of the core of a massive star which has worked its way up the fusion ladder fusing hydrogen to helium, helium to carbon, carbon to silicon and silicon to iron, synthesizing other elements as well along the way. When the mass of the iron core builds up to the Chandresekhar limit in a time scale of ~1 day, in a split second, it implodes and then explodes, blowing off the hydrogen-rich outer material as a supernova remnant, while the residual central material collapses to form a neutron star or, in the case of the most massive stars, a black hole.

A Type II supernova spectrum is shown in Fig. 13.3-1b. It also is unlike a blackbody spectrum and noticeably different from the Type Ia spectrum above it. The hydrogen Balmer series is prominent, and the lines, particularly the Hα line, show absorption-emission features. The blue-shifted absorption line is due to cooler hydrogen gas, seen in front of the hotter gas behind it, which is approaching us, while the unshifted emission line is due to hydrogen gas moving transversely to the line of site. The mechanism is the same as for the stellar winds of the hot O and B stars depicted in Fig. 12.4-2. The maximum expulsion velocity may be estimated from the difference in log λ between the start of the absorption line and the peak of the emission line (see Problem 13-2). In principle, the cosmological redshift of the supernova can be estimated from the displacement of the hydrogen emission peaks relative to the Balmer series rest wavelengths. However, the Type II supernovae exhibit much greater variability in their intrinsic brightnesses and are not useful as standard candles.

Overall, although they arise from totally different stellar systems, the numbers of Type Ia and Type II supernovae observed in external galaxies are approximately equal. We now know that the Chinese "guest star" was a Type II supernova, which left the Crab nebula and a neutron star at the center, and that Tycho Brahe witnessed a Type Ia supernova in 1572.

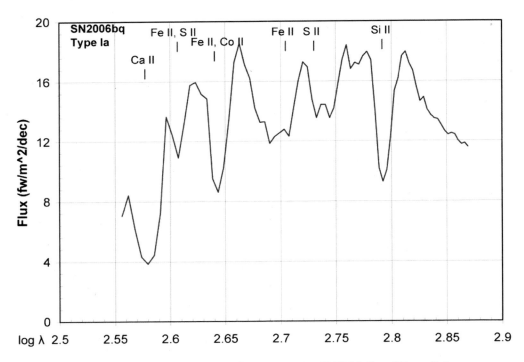

Fig. 13.3-1a. Spectrum of Supernova SN2006bq (Type Ia).

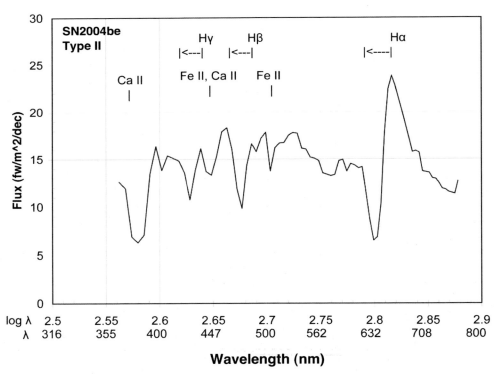

Fig. 13.3-1b. Spectrum of Supernova SN2004be (Type II).

Questions

13-1. What is the range of masses of stars which will eventually become white dwarfs, neutron stars, black holes?

13-2. What is the helium flash?

13-3. What two factors besides mass can have an effect on a star's fate?

13-4. What three types of pulsating stars are used as distance indicators?

13-5. What two factors make it more difficult to use the Mira variables as distance indicators?

13-6. Why are most novae and supernovae which occur in the Milky Way invisible to us?

13-7. Explain how supernova spectra can contain both blueshifts and redshifts of the spectral lines.

13-8. Why do some white-dwarf binary systems produce novae and others produce Type Ia supernovae?

13-9. What procedure is used to correct for variations of brightness in Type Ia supernovae to improve their accuracy as standard candles?

13-10. How are the combination absorption–emission lines in Type II supernovae produced?

Problems

13-1. Use the values of log L and log M in Table 13.2-1 to determine the slope and intercept of the mass–luminosity relation for Type I Cepheids by linear regression.

13-2. The spectrum of a Type II supernova in Fig. 13.3-1b has an Hα absorption–emission line with a spread between the beginning of the absorption and the peak of the emission of $\Delta\log\lambda = .025$. Use Eq. 10.3-2 to estimate the velocity of the expelled gas.

Chapter 14: Stellar Remnants

Compared to the densities of the stars that produce them, stellar remnants have extreme values of density. Central remnants are very dense. In white dwarfs, a mass comparable to the Sun's is packed into a star one hundred times smaller, so that the density is one million times as large. In neutron stars, a somewhat higher mass is even more tightly packed into a star nearly one thousand times smaller still, so that the density is one quadrillion times the Sun's. In black holes, a mass of more than three solar masses shrinks to a point, so that the density is infinite. In contrast, expanding remnants are very undense. In both planetary nebulae and supernova remnants, a mass ranging from a fraction of a solar mass to several solar masses is spread out over distances which grow from a few tens to hundreds of light years. They enrich the galaxy's star-forming gas in the heavier elements so that the next generation of stars has higher metal content.

14.1 White Dwarfs

White dwarfs are the end product of stellar evolution for stars in the mass range .4–8 M_o, which comprises the vast majority of stars that have had time to proceed to the final stage of evolution since the universe began. White dwarfs in close binary systems (separation < 1 AU) may give rise to either novae or Type Ia supernovae as their companions evolve, as discussed in Chapter 13. Single white dwarfs or those in more widely separated binary systems (separation > several AU) simply cool down slowly. They consist mostly of carbon and lack the high temperature required to support carbon fusion.

In the list of the twenty nearest star systems given in Appendix A7, there are two white dwarfs, namely the companions of Sirius and Procyon. Their basic properties are listed in Table 14.1-1.

Table 14.1-1. The Nearest White Dwarfs.

Star	Mass (Mo)	Radius (Ro)	Luminosity (Lo)	Temp (K)
Sirius B	.978	.0084	.0024	25200
Procyon B	.6	.02	.00055	7740

Note that Sirius B has a larger mass but a smaller radius. This inverse relation between radius and mass for white dwarfs continues up to the Chandresekhar limit of 1.38 M_o, at which point the star collapses because the electron degeneracy pressure cannot sustain the overlying weight. Sirius B is a considerably hotter and younger star than Procyon B. It is thought to have originated from a main sequence B4 or B5 star having a mass of ~5 M_o. In its red giant phase, it may have transferred some material to its companion Sirius A and enriched its metal content. The metallicity of Sirius A is about double that of the Sun. Procyon B came from a less massive star.

14.2 Neutron Stars and Black Holes

Neutron stars and black holes are the end product of stellar evolution for stars with masses greater than 8 M_o. Neutron stars have masses in the range 1.4–3 M_o, i.e. from the Chandresekhar limit to the neutron star limit where the repulsive nuclear forces cannot sustain the overlying weight. Neutron stars are much more common; black holes are very rare. It is thought that only the most massive stars, perhaps those with masses greater than 25 M_o, have sufficient mass remaining in the core after the Type II supernova event to form black holes.

Because of the great increase in angular velocity resulting from conservation of angular momentum as the core shrinks in radius by several hundred times, neutron stars rotate very rapidly. Many have rotation periods with time scales of milliseconds. Their strong magnetic fields beam the emitted radiation preferentially along magnetic axes and the rapid rotation swings these beams periodically across the line of sight in the same manner as a lighthouse, providing the pulsar phenomenon. Hundreds of pulsars are now known. The energy required to accelerate electrons to high velocities and emit electromagnetic radiation must come from rotational energy, so the rotation of neutron stars gradually slows down and the rotation period lengthens.

Because single black holes are "black", i.e. do not emit any electromagnetic radiation, most of the good candidates for stellar-mass black holes have been detected in binary star systems, in particular, in X-ray eclipsing binaries. Material is transferred from a companion star to the black hole via an accretion disk and heats up and emits X-rays before it disappears into the event horizon of the black hole. As the black hole and companion star orbit each other, the black hole periodically disappears from view behind the companion and the X-ray source is eclipsed. The orbital period and orbital velocities of the binary components can be measured. From the spectral type of the companion its size is fairly well known. Thus, the semimajor axis of the orbit can be estimated. The combined mass can be determined from Kepler's third law and the individual masses from the relative velocities.

The masses of the good candidates for stellar-mass black holes in our galaxy are plotted versus orbital period in Fig. 14.2-1. These systems range in distance from 2,700 ly to 39,000 ly with a median value of 8,500 ly. The companion stars are typically K- or M-type stars with masses of less than one solar mass. Cyg X-1 is an unusual and interesting system. The mass of the black hole is estimated to be 16 solar masses and that of the companion, an O9.7Iab star, 31 solar masses. Before it evolved to produce the black hole, the original star must have had a mass exceeding 50 solar masses, a value which is close to the highest observed value for stars in our galaxy.

In compact–object binary systems, in which the transfer of material from the companion to the compact object gives rise to X-ray nova outbursts, neutron stars can be distinguished from black holes by the brightness of the outburst. The surface of the neutron star acts like

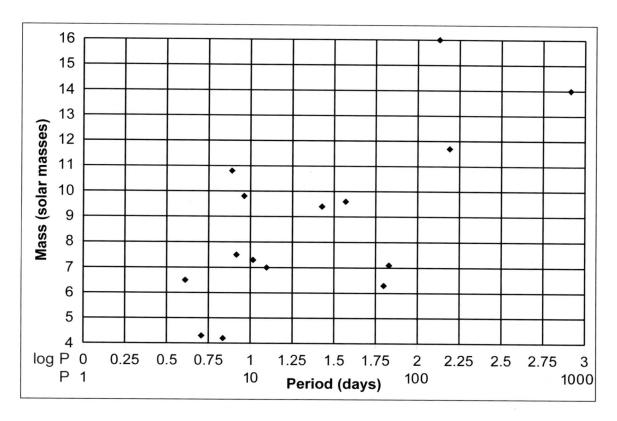

Fig. 14.2-1. Good Candidates for Stellar-Mass Black Holes.

a very reflective mirror, whereas the event horizon of the black hole is "fuzzy" and absorbs much of the radiation. As a result, X-ray novae involving a neutron star are typically a factor of 100 brighter than those involving black holes.

14.3 Supernova Remnants

Both Type Ia and Type II supernovae send out expanding clouds of gas known as supernova remnants. The mass of gas is limited to ~1.4 M_o in Type Ia events, but may exceed 20 M_O in the largest Type II events. The gas expanding at tens of thousands of km/s crashes into the interstellar gas, heating it and causing it to emit radiation over much of the electromagnetic spectrum. However, because these remnants are thousands of light years away, their ultraviolet, visible and even infrared radiation is strongly attenuated by the galactic dust and they are more easily observed in the radio and X-ray parts of the spectrum. Although less than 30 can be detected at visible wavelengths, more than 3000 are known in the radio region.

About 300 years ago a supernova occurred at a distance of 10,000 ly in the constellation Cassiopeia, producing the supernova remnant known as Cas A, which is not observed in the visible region but is the strongest such source at radio wavelengths and is relatively bright in the X-ray region as well. The radio spectrum of Cas A is shown in Fig. 14.3-1. The spectrum is featureless, showing only a steady decrease of flux toward longer wavelengths. Note the size of the total flux compared with the total flux of stars in the ultraviolet, visible, and infrared regions. Over the three-decade range shown, the total flux amounts to 2.0×10^{-4} nW/m^2, down a factor of several hundred thousand from the combined UV–VIS–IR flux of the brightest stars. That is why we need large radio telescopes to do radio astronomy!

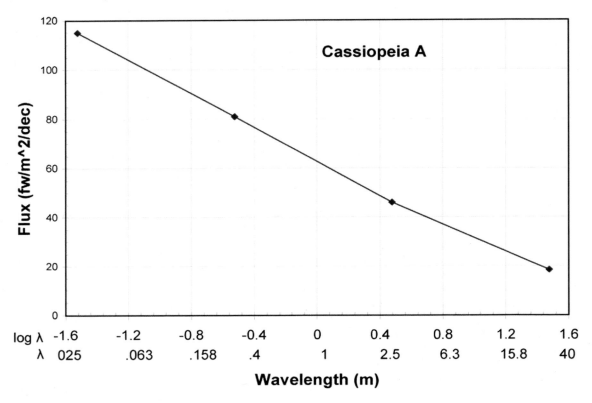

Fig. 14.3-1. Radio Spectrum of Cas A.

The X-ray spectrum of Cas A is shown in Fig. 14.3-2. Emission lines of several elements, produced in the supernova explosion and ejected at high velocity, are evident, the strongest being those of silicon and sulfur. The spectrum is an emission spectrum because the supernova remnant consists of a hot <u>transparent</u> gas, and resembles a nebula more than a star. Although the spectrum deviates significantly from a blackbody spectrum, a rough measure of the temperature of the gas may be determined by placing the blackbody spectrum over the X-ray spectrum, sliding it to attain a best fit and then determining λ_{max} and T from Wein's law (see Problem 13-7). The temperature is in the range of millions of Kelvins.

124   *Exploring the Universe*

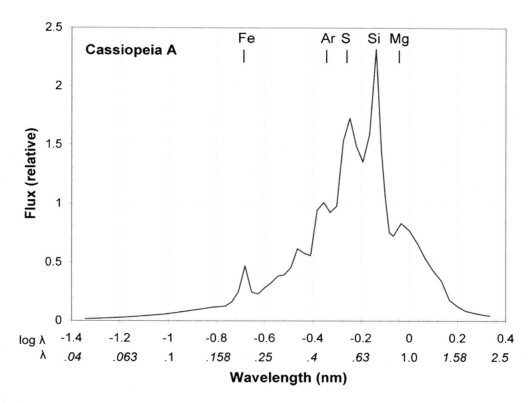

Fig. 14.3-2. X-Ray Spectrum of Cas A.

Questions

14-1. What is the fate of a single white dwarf?

14-2. How might Sirius A have become enriched in metals?

14-3. Why do neutron stars spin so rapidly?

14-4. What two methods can be used to tell whether a compact object is a neutron star or a black hole?

14-5. At what distance are the good candidates for stellar-mass black holes located?

14-6. Is the fact that the companion stars of good candidates for black holes are low-mass K- or M-type stars compatible with what we know about the relative numbers of different types of stars?

14-7. Why are supernova remnants best observed in the radio and X-ray parts of the spectrum?

14-8. How do bright radio sources such as Cas A compare in total flux with bright UV–VIS–IR stars such as Sirius?

Problems

14-1. Suppose a white dwarf had the same radius as Procyon B but a temperature twice as hot. What would be its luminosity in units of L_o?

14-2. Place the blackbody spectrum over the X-ray spectrum of Cas A and slide it to attain a best fit. Determine λ_{max} and T from Wien's law.

Chapter 15: The Milky Way

Although astronomers have realized for a long time that we live in a disk-shaped star system, the determination of the exact shape of this system and of our position in it has been difficult for two reasons. First, we are inside the system and cannot get an overall view of it. Second, interstellar dust strongly attenuates starlight so that our view in the galactic plane is very limited in distance by an amount which is highly variable in direction. Determination of our galaxy's shape and of our position in it has made use of observations for which the interstellar attenuation is reduced, either by observing objects away from the galactic plane or by observing in the radio and infrared parts of the spectrum.

15.1 Our Position in the Milky Way

In the late eighteenth century, William Herschel counted the number of stars visible in different directions in the sky and concluded correctly that the galactic disk is at least several times wider than it is thick, and incorrectly that we are near its center. Because interstellar attenuation permits a view which extends only to a small fraction of the entire size of the disk, conclusions about our position based on star counting are erroneous.

In the early twentieth century, the Dutch astronomer Jacobus Kapteyn used the proper motions of stars to infer their distances and gauge the size of the Milky Way. Essentially, he was using Eq. 2.3-2 and assuming that the average random speeds of stars are the same at different distances so that the distance is inversely proportional to average angular speed. Although the method is valid in principle, Kapteyn did not understand the severity of interstellar attenuation, which prevented him from seeing the more distant stars. His value of 20 kpc for the diameter of the Milky Way was too small by a factor of two.

Soon after Kapteyn's work, the American astronomer Harlow Shapley used the RR Lyrae variables in ~70 globular clusters as distance indicators to determine the cluster's distances and produce a three-dimensional plot of their positions in space. He correctly assumed that the center of this distribution would be close to the center of the Milky Way, and clearly showed that the Sun's position is displaced significantly from the center (two-thirds of the way out in Shapley's initial result, halfway out according to modern measurement). As with Kapteyn, the lack of knowledge of interstellar attenuation affected Shapley's result for the size of the Milky Way, in this case increasing its estimated size. Because Shapley failed to account for the attenuation properly, the distances of the further globular clusters were too large. In addition, he did not understand the difference in the calibration between the RR Lyrae and Cepheid variables. As a result, his value of 300,000 ly for the diameter of the Milky Way was almost three times too large. Later, when astronomers had a more accurate understanding of both interstellar attenuation and the RR Lyrae calibration, this estimate was revised downward to close to the modern estimate of 100,000 ly. However, the estimate of the sizes of external galaxies remained in error until the distinction between Type I and Type II Cepheids was realized in the 1950s.

There are now about 160 globular clusters known, which are satellites of the Milky Way. In using their three-dimensional spatial distribution to locate the Sun's position in the Milky Way, one must be aware that they separate into two groups according to their metallicity, i.e. their relative abundances of elements heavier than hydrogen and helium. Quantitatively, the metallicity is equal to the logarithm of the quotient of the cluster's iron/hydrogen abundance ratio and the Sun's iron/hydrogen abundance ratio:

$$\text{Metallicity} = \log [\text{Fe/H ratio for cluster} / \text{Fe/H ratio for Sun}]$$

Thus, a metallicity of 0 means a metal content equal to the Sun's, a metallicity of –1 means a metal content equal to 10% of the Sun's, etc. The distribution of metallicities for the Milky Way's globular clusters is shown in Fig. 15.1-1. The distribution is clearly bimodal: A low-metallicity group, with metallicities of less than –.8 comprises ~75% the clusters, while a high-metallicity group, with metallicities of greater than –.8, comprises the other ~25%. Clusters in the low-metallicity group are generally older and their spatial distribution is more spherical, reflecting the conditions which prevailed in the early universe. Clusters in the high metallicity group are generally younger and their spatial distribution is more flattened, reflecting the enrichment of the heavy elements and flattening of the galactic shape with time. Note how the proportion of clusters with at least four identified RR Lyrae variables (darker portion of the histogram bars) is zero or very small for the high-metallicity clusters and ~50% or more for the low-metallicity clusters.

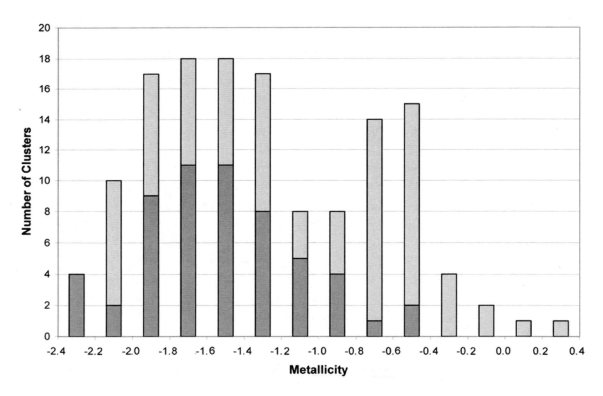

Fig. 15.1-1. Distribution of Metallicity for the Milky Way's Globular Clusters.

Part of Laboratory Exercise LE4 is to determine the Sun's position in the Milky Way by using the distances to 57 globular clusters which contain at least four identified RR Lyrae variables. The current accepted value for the distance of the Sun from the center is 8.5 kpc or 27,500 ly.

15.2 The Center of the Milky Way

The density of the stars at the center of the Milky Way is truly astounding. A close-up image of the central few light years taken in infrared light at a wavelength of 2200 nm shows hundreds of stars, whereas there would be no stars at all in our neighborhood. Observations over several years have revealed that several bright red-giant stars are orbiting about a very massive object in the center. From the estimated period of the orbit and the calculated semimajor axis, the mass of the central object may be determined (See Problem 15-2). It is most likely a supermassive black hole which has grown by accretion of gas from the dense central region falling into it.

A wider view of the central region in infrared light indicates the presence of a bar in the galactic structure so that the Milky Way is now considered to be a barred spiral of type SBb. In a recent computer model of the structure of spiral galaxies, bars come and go as gas keeps infalling into the galaxy and creating new stars. In this scenario, the Milky Way is now in one of its barred phases.

15.3 The Orbit of the Sun in the Milky Way

In response to the gravitational pull of the matter in the Milky Way galaxy, the Sun and its neighboring stars orbit the center at a speed of ~220 km/s. How is this speed determined? It must be determined by measuring the average Doppler shift with respect to a set of objects out of the galactic disk, whose average velocity with respect to the galactic center is zero. Two such sets of objects are the low-metallicity globular clusters and nearby galaxies. (High-metallicity globular clusters have a rotational component because they formed later out of material which was participating in the disk rotation.)

It is believed that the orbits of stars in spiral galaxies are neither perfectly circular nor elliptical. The star attains a closest point to the center and a farthest point from it, but these rotate appreciably in angle during one orbit because, unlike the case of planets orbiting the Sun, the attractive gravitational force does not vary inversely as the square of the distance from the center. However, for simplicity, we may approximate the orbit as a circle.

<u>Example Problem 15.2-1:</u> Determine the orbital period of the Sun around the galactic center.

$$P = 2\pi r/v$$
$$r = 2.75 \times 10^4 \text{ ly} \times 9.4605 \times 10^{12} \text{ km/ly} = 2.60 \times 10^{17} \text{ km}$$

$$P = 2\pi \times 2.60 \times 10^{17} \text{ km} / (2.2 \times 10^2 \text{ km/s})$$
$$P = 7.42 \times 10^{15}\text{s}$$

$$\boxed{P = 2.4 \times 10^8 \text{ yr} = 240 \text{ million yr}}$$

In principle, the diameter of the Sun's orbit around the galactic center could be used as a baseline to achieve the ultimate parallax measurement and determine very accurate distances to other galaxies and to the farthest reaches of the visible universe. From Eq. 2.8-1, the parallax p, in radians, of an object at distance d, in light years is given by

$$p = D/2d \qquad\qquad\qquad\qquad \text{(Eq. 15.3-1)}$$

where the diameter of the orbit $D = 5.5 \times 10^4$ ly.

Example Problem 15.3: Using Eq. 15.3-1, calculate the parallax of the Andromeda galaxy at a distance of 2.9×10^6 ly.

$$p = 5.5 \times 10^4 \text{ ly} / (2 \times 2.9 \times 10^6 \text{ ly})$$
$$p = 9.5 \times 10^{-3} \text{ radians}$$

$$\boxed{p = 1960 \text{ arc sec.}}$$

This is a little larger than the diameter of the full moon.

It is left as a problem (Problem 15-4) to show that, using the diameter of the Sun's orbit as a baseline, the parallaxes of the apparently most distant objects in the visible universe are still larger than the parallaxes of the nearest stars using the diameter of the Earth's orbit. This comparison makes an important point: intergalactic distances, relative to the sizes of galaxies, are far smaller than interstellar distances (in our stellar neighborhood), relative to the sizes of earth-like planetary orbits.

Of course, waiting 240 million years for the Sun to complete its orbit is impractical. However, maybe we do not have to wait so long. If there is a Galactic Internet, with different views of the universe taken from different parts of the galaxy, then maybe such accurate parallaxes already exist. All we have to do is find the right password!

Questions

15-1. Why cannot star counting be used to determine the Sun's position in the Milky Way?

15-2. Both Kapteyn and Shapley failed to account adequately for interstellar attenuation but Kapteyn got a low value for the size of the Milky Way and Shapley got a high value. Why?

15-3. Who was the first astronomer to show that the Sun was not at the center of the Milky Way?

15-4. In what ways does the low-metallicity group of globular clusters differ from the high-metallicity group?

15-5. How is the mass of the massive object at the center of the Milky Way determined?

15-6. What is the galactic classification of the Milky Way (Sa, SBa, etc)? How do we know this?

15-7. If we could use our galactic orbit as a baseline, why would galactic parallaxes be so much larger than stellar parallaxes using the Earth's orbit as a baseline?

15-8. How is the speed of the Sun's orbital motion around the galactic center determined?

Problems

15-1. Using a = 1000 AU and P = 20 years, and neglecting the second mass M_2, solve for the mass of the black hole at the center of the Milky Way. Use $M_1 + M_2 = a^3/P^2$.

15-2. Calculate the Schwarzschild radius of this black hole using R (km) = 3 × M (solar masses).

15-3. The spiral galaxy M 100 in the Virgo cluster is located at a distance of 56 million ly. Using the diameter of the Sun's orbit as a baseline, calculate its parallax.

15-4. As in Problem 3, calculate the parallax of the apparently most distant objects visible to us (d = 5.76 billion ly).

Chapter 16: Galaxies

It was as recently as the 1920s that astronomers realized that our own Milky Way galaxy does not comprise the universe, but that the universe is populated by trillions of such star systems, some hundreds of times as massive as the Milky Way and some thousands of times as luminous. Bright galaxies are of two major types: spirals and ellipticals. Galaxies are grouped into clusters and the clusters into superclusters, forming a vast filamentary structure with enormous voids between the filaments. Most galaxies are luminous because of their component stars, but some emit enormous energy from active regions in their cores. The brightest of these can be detected at great distances and permit us to glimpse the universe when it was very young.

16.1 Normal Galaxies

In the early 1920s, there was a great debate in the astronomical community concerning the spiral nebulae, which astronomers had observed for decades, but whose exact nature remained unclear. Harlow Shapley championed the idea that the spiral nebulae were part of the Milky Way, while Heber Curtis championed the opposing view that they were external star systems like the Milky Way. Curtis advanced three arguments to support his view:

1. The spiral nebulae exhibit a wide range of angular sizes, from a few arc seconds to 3° for M 31 in Andromeda – a factor of several thousand. This wide range is most naturally explained by a large range of distances rather than by a large range of true sizes.
2. They are not observed in a region called the <u>zone of avoidance</u> in the plane of the disk of the Milky Way. If they were in the Milky Way, one would expect to see some close ones in the disk plane, although the more distant ones might be obscured.
3. Their spectra are absorption spectra like those of star clusters, not emission spectra like those of true nebulae such as the Orion nebula.

The debate was settled fairly quickly in 1924 when Edwin Hubble observed Cepheid Variables in M 31 and derived a distance to it of half a million light years. Due to an underestimation of the true brightness of the Cepheids as well as some other errors, Hubble's original value of the distance was far too low, but even half a million light years put it well outside the Milky Way.

Observations of galaxies soon revealed that the bright galaxies are of two major types, namely spirals and ellipticals, with the spirals being more prevalent. Hubble placed the various types on a diagram which became known as the <u>tuning-fork diagram</u>, which showed the normal spirals (S) and barred spirals (SB) ranging from tightly wound (Sa, SBa) to loosely wound (Sd, SBd), and the ellipticals ranging from spherical (E0) to very flattened (E9). Hubble never intended that the diagram be interpreted in an evolutionary sense, although there were some attempts to do so involving evolution from an irregular

(I) shape through a loose spiral shape and then a tight spiral shape to an elliptical shape. The modern view is that the form of an <u>isolated</u> galaxy remains pretty much fixed, and is determined by the relative amounts of organized rotational motion versus random motion: Strong rotation tends to produce spirals and strong random motion ellipticals. However, evolutionary changes of shape do occur in dense galactic clusters through <u>collisions</u>. The planes of rotation of two colliding spirals are oriented randomly and the resulting galaxy will have more random motion and be more elliptical. The giant elliptical galaxies often seen at the centers of dense clusters, such as M 87 in the Virgo cluster, are probably the result of many such collisions and mergers.

Hubble also studied the radial velocities of external galaxies, i.e. the component of velocity toward or away from us, and showed that, on average, the galaxies are receding from us at a velocity which is proportional to their distance in a relation which came to be known as Hubble's Law:

$$v = Hd \hspace{3cm} \text{(Eq. 16.1-1)}$$

where v is usually measured in km/s and d in megaparsecs. Due to an underestimate of the true brightness of the Cepheid variables as well as some other errors, Hubble's original value for the Hubble constant H of 500 km/s/Mpc was much too large. It is now believed that the true value of H lies close to 70 km/s/Mpc.

The recession velocity of a galaxy is determined by measuring its <u>redshift</u>, i.e. the proportional shift of the wavelengths of the lines in its spectrum. The redshift z is defined as

$$z = \lambda'/\lambda - 1 \hspace{3cm} \text{(Eq. 16.1-2)}$$

where λ' is the observed wavelength and λ is the rest wavelength. If spectra are displayed on a logarithmic wavelength scale, and a standard spectrum slid along the log λ axis until the spectral lines coincide, giving a displacement of $\Delta\log \lambda$, the wavelength ratio z is given by

$$\lambda'/\lambda = 10^{\wedge}\Delta\log \lambda \hspace{3cm} \text{(Eq. 16.1-3)}$$

so that the redshift z is given by

$$z = 10^{\wedge}\Delta\log \lambda - 1 \hspace{3cm} \text{(Eq. 16.1-4)}$$

At velocities which are small compared with the velocity of light, i.e. v/c $\ll$ 1, the redshift z is approximately equal to v/c so that we may use

$$v/c = z \hspace{3cm} \text{(Eq. 16.1-5)}$$

At larger velocities, the full expression from Einstein's theory of relativity must be used

$$v/c = [(1 + z)^2 - 1] / [(1 + z)^2 + 1] \hspace{2cm} \text{(Eq. 16.1-6)}$$

<u>Example Problem 16.1-1:</u> The spectral lines in a galaxy's spectrum are displaced by an amount $\Delta \log \lambda = .05$. Calculate its redshift z, its recession velocity v, and its distance d, using $H = 70$ km/s/Mpc.

$$z = 10^{\wedge} \Delta \log \lambda - 1 = 10^{\wedge}.05 - 1 = 1.122 - 1$$

$$\boxed{z = .122}$$

$$v = z \times c = .122 \times 3 \times 10^5 \text{ km/s}$$

$$\boxed{v = 3.66 \times 10^4 \text{ km/s}}$$

$$d = v/H = (3.66 \times 10^4 \text{ km/s}) / (70 \text{ km/s/Mpc})$$

$$\boxed{d = 523 \text{ Mpc.}}$$

By the 1930s, astronomers had observed clusters of galaxies containing hundreds and thousands of members. Measurements of the velocities of the galaxies within the clusters showed that they were moving too rapidly for the cluster to hold together if the only mass in the cluster were the mass associated with the luminous stars. These anomalously high velocities were the first evidence of <u>dark matter</u>, whose presence has since been indicated in many other ways and whose exact nature remains the subject of many investigations.

With the advent of computer-controlled telescopes in the 1980s and the automatic recording of redshifts of thousands of galaxies, studies of the large-scale structure of the universe become possible. These were usually presented as two-dimensional maps of the galaxies within a thin wedge-shaped volume of the sky ~120° in angular width out to distances of .5 – 2 billion ly. They revealed that the galaxies and clusters were arranged in long filamentary superclusters with giant voids having sizes of 50 – 250 million ly in between. This structure extends in all directions and, presumably, throughout the universe, and it appears that the voids are limited to a size of about 300 million ly, so that on a scale of ~1 billion ly or greater the universe is homogeneous. Note that conclusions about such structure are independent of the value of the Hubble constant H. A different value of H would change the absolute values of the galaxies' distances but not their relative distances, and would not distort the structure.

Images of the galaxies taken in different spectral regions have yielded additional information about their structure and dynamics as well as shown them in their total beauty. This extension of our vision beyond the visible region is illustrated by the composite image of the Andromeda galaxy M 31 on the front cover of this book. This image was created by combining three images:
1. A far ultraviolet image in a band centered at 155 nm which is a composite image from the Galex mission.
2. A visible image in the region 400 – 700 nm taken by John Corban.
3. An infrared image in the J band centered at 1250 nm from the 2 Mass archive.

The combined image was created by displaying the UV in blue, the visible in green, and the IR in red and adjusting the balance of the three colors until a desirable result was achieved.

The ultraviolet component comes mainly from the spiral arms and rings, where many O and B stars are located. The visible component is strongest from the nuclear bulge, but also shows the spiral arms clearly. The infrared component is generally amorphous and concentrated near the nuclear bulge. However, unlike the ultraviolet and the visible, the infrared penetrates the dust so that the dust lanes look dark red instead of totally black.

This image is displayed in an inverted position from the usual position shown for M 31 in astronomy textbooks because, in this inverted orientation, our visual system receives unambiguous clues from all components that we are looking down on the galaxy from above. In the normal orientation, the ultraviolet component (blue) indicates that we are looking down from above but the infrared component (red) indicates that we are looking up from below, yielding an ambiguous image. Turn your book upside down to see the effect. This effect probably results from the fact that, when there is no strong indication of the direction of view, as there is not in the ultraviolet (blue) component, our visual system defaults to the usual direction, which is down from above on most occasions in everyday life.

16.2 Active Galaxies

With the advent of radio astronomy in the 1940s, it became evident that a small but important minority of galaxies emit unusual amounts of energy in the radio region and other parts of the electromagnetic spectrum, which cannot be explained as resulting from a group of normal stars. Over several decades, these active galaxies were catalogued into four basic groups, whose principle characteristics are listed in Table 16.2-1.

Table 16.2-1. Characteristics of Active Galaxies.

Object	Seyfert Galaxies	Radio Galaxies	Quasars	BL Lac Objects
Galaxy Type	Spiral	Elliptical	Spiral, Elliptical	Elliptical
Brightest λ of Nucleus	Infrared	Radio, X-ray	Infrared	Infrared
Emission Lines	Iron & Metals	–	Hydrogen	Weak
Brightest λ of Jets & Lobes	–	Radio, X-ray	Radio	Radio
Luminosity (rel. to M.W.)	.1 – 10	.1–10	10–10^4	10^3–10^5
Range of Redshift z	Small $\rightarrow$	Small $\rightarrow$	.06 $\rightarrow$ 5	.06 $\rightarrow$ 5
Timescale of Variability	Weeks $\rightarrow$ Months	–	Days $\rightarrow$ Months	Days $\rightarrow$ Months
Year of Discovery	1943	1944	1960	1929*

* At first mistaken for a variable star; true nature recognized later.

Seyfert galaxies and radio galaxies are distinguished from quasars and BL Lac Objects by luminosity and by distance, the latter being distinguished from radio galaxies by the type of galaxy (spiral versus elliptical) and by the brightest wavelength range of emission (infrared versus radio or X-ray). BL Lac objects are distinguished from quasars by their much weaker spectral lines and greater luminosity.

Since the late 1960s, astronomers have attempted to create a unified model of active galaxies based on a supermassive black hole at the center. Gas and perhaps even entire stars fall into the black hole, transferring their gravitational potential energy into kinetic energy and some material is expelled at high speeds, even approaching the speed of light, in both directions along the magnetic axis. Because of the high directionality of the emitted material and radiation, the observed luminosity depends strongly on our line of sight relative to the magnetic axis. If we are viewing the galaxy along the magnetic axis, we see a BL Lac object; if we are viewing close to it, we see a quasar; if we are viewing farther from it, we see a Seyfert galaxy or radio galaxy. Observations of stellar velocities near the center of the Milky Way and other galaxies has revealed that supermassive black holes do exist at the center of many, and perhaps even all, large star systems and this discovery has placed the supermassive-black-hole model on a firmer footing.

However, the angle of view with respect to the magnetic axis cannot be the entire story. The fact that we observe no nearby quasars or BL Lac objects shows that evolutionary effects are involved. Evidently, the conditions which produce this very luminous activity no longer exist in present day galaxies. It appears that quasar-like activity is a phase which galaxies, maybe including the Milky Way, go through when they are relatively young compared to the current age of the universe.

Figure 16.3-1 shows a composite quasar spectrum, which is an average of the spectra of many quasars. The strong emission lines in the spectrum are due to hydrogen (the Lyman α line) and ionized forms of carbon, oxygen, silicon, and magnesium. The wavelengths are the rest wavelengths and this spectrum, also provided on a transparency, can be placed over individual quasar spectra to determine their redshifts. Figures 16.3-2a and 16.3-2b show spectra of two quasars, in which the wavelengths are the observed wavelengths. Part of Laboratory Exercise LE5 is to determine the redshifts and the recession velocities of these quasars.

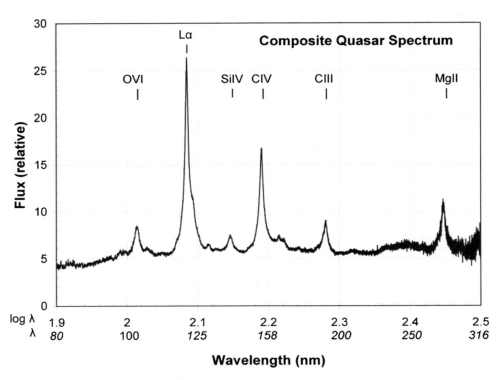

Fig. 16.2-1. Composite Quasar Spectrum.

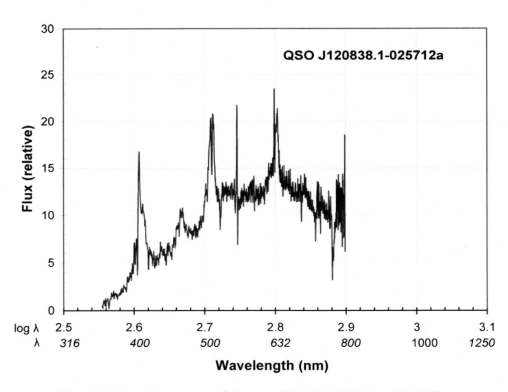

Fig. 16.2-2a. Spectrum of Quasar QSOJ120838.1 − 025712a.

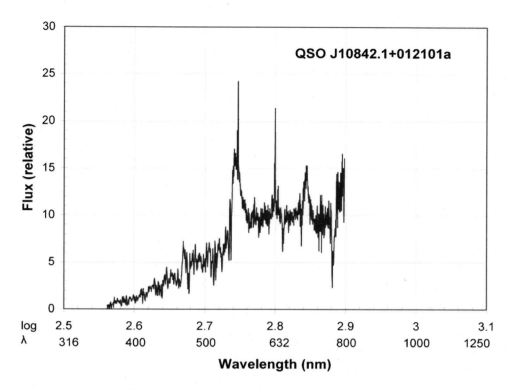

Fig. 16.2-2b. Spectrum of Quasar QSOJ10842.1 + 012101a.

Questions

16-1. What three arguments did Heber Curtis advance to support the claim that the spiral nebulae were external galaxies?

16-2. Under what conditions do spiral galaxies evolve into ellipticals?

16-3. Why was Hubble's original value of the Hubble constant too large?

16-4. What was the first evidence of dark matter?

16-5. Describe the large-scale structure of the universe.

16-6. What are the prominent features in ultraviolet and infrared images of spiral galaxies?

16-7. What are the distinguishing features of Seyfert galaxies and radio galaxies?

16-8. What are the distinguishing features of quasars and BL Lac objects?

16-9. Describe the unified model which has been proposed to explain active galaxies?

16-10. Can the differences among active galaxies all be explained by different angles of view with respect to their magnetic axes? Explain.

Problems

16-1. A distant quasar has a spectrum in which the Lyman alpha line at $\lambda = 121.53$ nm is red-shifted to the Balmer alpha position at $\lambda = 656.3$ nm. Calculate its redshift z, its recession velocity v, and its distance d, using $H = 70$ km/s/Mpc.

16-2. In the above quasar spectrum, what are the shifted wavelengths of the Lyman β and γ lines? Do they coincide with the Balmer β and γ positions? Consult Table 10.2-1.

Chapter 17: Cosmology

Two major questions concerning the universe in which we live, which have puzzled astronomers for centuries, are: Where did we come from? and, Where are we going? Finally, in the later half of the twentieth century, astronomers began to collect data which has helped to answer these questions quantitatively. Information about the origin of the universe is contained in the cosmic background radiation, whose spectrum proves the existence of an early hot dense fireball, and whose slight non-uniformities reflect the slight density variations which grew to be the current large-scale cosmic structure. Information about the fate of the universe is contained in careful measurement of redshift versus distance, and indicates that the universe will continue to expand forever at an increasing rate.

17.1 The Cosmic Background Radiation

The cosmic background radiation was first detected in 1965 by Arno Penzias and Robert Wilson, accidentally, as they were testing a microwave receiver at Bell Laboratories. It was soon realized that this all-sky microwave radiation was the observable remnant of the cosmic explosion known as the Big Bang, which began our universe some 13.6 billion years ago. As the universe cooled enough for neutral hydrogen atoms to form out of individual protons and electrons, matter quickly changed from opaque to transparent, and the radiation in equilibrium with matter at the time was free to propagate through space unimpeded. The spectrum of the cosmic background radiation is shown in Fig. 17.1-1. It can be very closely fit by a blackbody spectrum having a temperature of 2.73 K, as can be seen by placing the blackbody spectrum transparency on it and sliding it to attain a best fit.

Although the radiation is often referred to as the 2.73 K cosmic background radiation, it is more accurate to think of the shift of the spectrum from the visible region into the microwave region ($\lambda_{max} = 1.35$ mm) as a redshift effect due to the expansion of the universe rather than a cooling effect. There was not or is not any matter at a temperature of 2.73 K interacting with this radiation. It was emitted from matter at a temperature of ~3000 K and initially had much shorter wavelengths, but those wavelengths have been stretched out over billions of years by the stretching of space associated with the expansion of the universe. This spectrum is the most highly redshifted one (z ~1100) which we see or will ever see.

The cosmic background radiation is also extremely bright. From Eq. 10.1-7, the total flux from a blackbody source at temperature T covering the entire sky (solid angle $\Omega = 4\pi$) is given by

$$F = 4\sigma T^4 \qquad \text{(Eq. 17.1-1)}$$

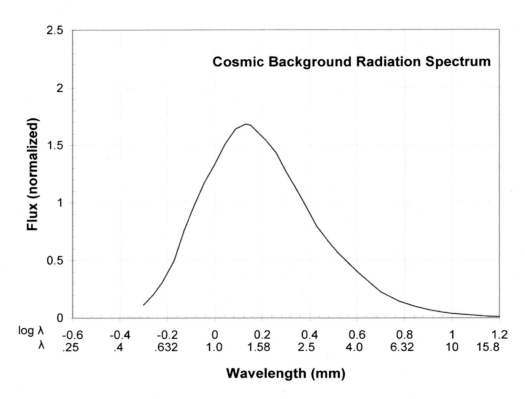

Fig. 17.1-1. The Cosmic Background Radiation Spectrum.

Example Problem 17.1-1: Calculate the total flux and combined magnitude of the cosmic background radiation.

$$F = 4\sigma T^4$$
$$F = 4 \times 5.6705 \times 10^{-8} \text{ W/m}^2\text{/K}^4 \times (2.73 \text{ K})^4$$

$$\boxed{F = 1.26 \times 10^{-5} \text{ W/m}^2}$$

$$m = -2.5 \log (F/(25 \times 10^{-9} \text{ W/m}^2))$$

$$\boxed{m = -6.8}$$

This is brighter than any other steady celestial source (besides the Sun and the Moon) in any wavelength range. The cosmic background radiation also contains a large number of photons. The density of microwave-radiation photons is 4×10^8 /m^3 while the density of hydrogen atoms is 1/(3 m^3) – a ratio of 1.2 billion to one! Unmistakable evidence of the early universe is still very much with us.

Non-uniformities in the cosmic background radiation were first detected by the Cosmic Background Explorer (COBE) satellite in the early 1990s and imaged with much better resolution by the Wilkinson Microwave Anisotropy Probe (WMAP) in the early 2000s. The non-uniformities are on the order of 1 part in 10^5, typically ~1° in angular size, and reflect slight differences in density at the time the universe changed from opaque to transparent ~380,000 years ago. Through gravitational attraction, these small differences amplified through eons of time to become the large-scale structure of superclusters and voids which we see today. Using a cosmological model which contains ordinary matter, dark matter, and dark energy (required to explain the accelerating expansion discussed in the next section), the WMAP data places fairly tight constraints on the amounts of the three components and on the geometry and age of the universe.

Table 17.1-1. Composition, Geometry, and Age of the Universe.

Composition	
Ordinary Matter	4.2% ± 0.3%
Dark Matter	23 ± 3.0%
Dark Energy	72 ± 3%
Geometry	flat
Age (billion years)	13.7 ± 0.4

17.2 Acceleration of the Expansion of the Universe

For many decades following the realization in the 1920s that the universe is expanding, it seemed that there are two possible fates:

1. The universe will continue to expand forever by a continuously decreasing rate as gravitational attraction between its constituent masses slows the expansion rate.
2. The universe will reach a maximum size and then begin a contraction phase, leading ultimately to a "Big Crunch" and possibly to another Big Bang.

In order to decide between these alternatives, one must determine the rate of deceleration of the expansion: a slower rate would yield the first scenario and a faster rate the second. This measurement is accomplished by following the redshift versus distance relation, first plotted by Hubble, to large redshifts and distances. Accurate determination of distance requires the use of a very bright "standard candle".

The brightest Cepheid variables can only be seen out to distances of ~100 million light years, which is not nearly far enough to gauge accurately a change in the expansion rate. In the late 1990s, studies of the Type Ia supernovae showed that, if they were properly calibrated by observing their light curves (brighter Type Ia supernovae decline more slowly), they could be used out to much greater distances of ~ several billion light years. Measurements by several groups consistently showed that, at redshifts greater than ~0.3, the supernovae appeared significantly dimmer, i.e. at greater distances than expected.

Thus, at this observed time in the past, the Hubble parameter H, which is the slope of the recession velocity versus distance curve was smaller and has been increasing since. Therefore, the rate of expansion has been increasing. This <u>acceleration</u> of the expansion rate cannot be explained by the presence of matter, even dark matter, since the attractive gravitational force of any matter would act to slow the expansion rate. In order to describe the observed behavior of the expansion rate, a term arising from a component called <u>dark energy</u> must be introduced into the cosmological equations. Although very imperfectly understood, this dark energy is now included in all cosmological models and comprises nearly three quarters of the matter–energy components of our universe.

Recent measurements have shown that at even higher redshifts ($z \sim 1.2$) the redshift versus distance curve turns around again and exhibits <u>deceleration</u> of the expansion rate. The picture that emerges is the following. Early in its history, the universe is dense and the change in the expansion rate is dominated by the <u>decelerating</u> effect of the <u>attractive</u> gravitational force. As the density thins out the <u>accelerating</u> effect of the <u>repulsive</u> dark energy component comes to dominate, insuring that the expansion will continue forever.

17.3 Distances in Cosmology

In the everyday world, in which objects are either static or have speeds small compared with the speed of light, we deal with a single distance to other objects. In cosmology, distant objects can have recession velocities approaching the speed of light and one must take care in defining distance. There are four principal distances which are used in cosmology. These are listed in Table 17.3-1 in order of size, from largest to smallest.

Table 17.3-1. The Four Distances Used in Cosmology.

Name	Symbol	Definition
Luminosity distance	d_L	Distance at which a standard candle has the observed apparent brightness
Now distance	d_{now}	Distance measured by a chain of observers along the line of sight
Light-travel-time distance	d_{ltt}	Distance which light has covered since being emitted by an object
Angular-size distance	d_{as}	Distance at which a standard yardstick has the observed angular size

The luminosity distance is the one that we determine using a standard candle such as the Type Ia supernovae. The now distance is the sum of incremental distances measured by a chain of observers along the line of sight. It exists in principle but cannot be observed directly because light (and information) cannot travel to us any faster than the speed of light c. The light-travel-time distance is the distance covered by light emitted from an object at time t_{em} and received by us at a time t_o, and is equal to $c(t_o - t_{em})$. This is the distance that we naturally associate with a length of time in the past at which an object appears. The angular-size distance is the distance determined from Eq. 2.2-1, using a standard value of the size D and the measured value of angular size α. It is really the best measure of the apparent distance of an object.

In general, the exact values of these distances depend on the cosmological model used, i.e. the amounts of ordinary matter, dark matter, and dark energy. However, there are two relations between them which are independent of the model: the first is

$$d_{as}(t_o) = d_{now}(t_{em}) \qquad \text{(Eq. 17.3-1)}$$

i.e. the angular-size distance measured at the current time is equal to the now distance at the time of emission. The second is

$$d_L = (1 + z)^2 d_{as} \qquad \text{(Eq. 17.3-2)}$$

This second relation may be understood as follows: If a blackbody source has a redshift z, its redshifted temperature T' will be $T/(1 + z)$ and its apparent brightness decreased by a factor of $(1 + z)^4$. In order to account for this decrease by the inverse square law, the luminosity distance must be $(1 + z)^2$ times as large. The current consensus is that the universe has a flat geometry and, in that case, an additional relation holds:

$$d_{now} = (1 + z)d_{as} \qquad \text{(Eq. 17.3-3)}$$

so that d_{as}, d_{now}, and d_L are all related by factors of $(1 + z)$.

The four distances are approximately equal for small values of z but diverge significantly for $z > 0.1$. While d_L, d_{now}, and d_{ltt} continue to increase with increasing z, d_{as} reaches maximum and then decreases again. For the current consensus values of the cosmological parameters, this turnaround occurs at $z = 1.645$ and $d_{as} = 5.76$ billion light years. The light seen from these objects was emitted at a time 4.01 billion years after the big bang. These are the <u>apparently most distant</u> objects we can see. Light now seen with a higher value of redshift, i.e. $z > 1.645$, was emitted at an earlier time by objects which were <u>closer</u> at that time but required a <u>longer time</u> to reach us because of the higher rate of expansion in the early universe.

The angular-size distance turnaround results in an angular-size turnaround of a standard yardstick, shown in Fig. 17.3-1 for a standard galaxy having a diameter of 100,000 light years. It reaches a minimum angular-size of 3.58 arc seconds at $z = 1.645$ and then increases beyond this point. This effect is good news for astronomers observing galaxies with high redshifts. They do not get so small that the angular resolution becomes very poor. The bad news is that they become very faint. The apparent brightness is inversely proportional to the square of the luminosity distance, which continues to increase steeply at this point.

Another effect that must be considered is <u>gravitational lensing</u>. Distant objects can often be magnified and brightened by the bending of their light by a closer cluster of galaxies along the line of sight.

The axes in Fig. 17.3-1 are significant because <u>redshift</u> and <u>angular-size</u> are two of the three principal directly measurable quantities of distant objects. The third is <u>apparent brightness</u>, and it is from the plot of apparent brightness of Type Ia supernovae versus redshift that we know about the accelerating expansion of the universe. The difficulty in using the plot of angular-size versus redshift is that it is more difficult to find a good standard yardstick than a good standard candle. Individual galaxies are carried along with the expansion of the universe, but do not expand themselves. It is generally believed that galaxies like the Milky Way have remained close to their present size since birth, but we do not know for sure that they have not increased in size by perhaps 25% or so due to the addition of new material. In addition, because of the faintness of high-redshift galaxies, it is difficult to discern their complete outline and get an accurate measure of the angular diameter. However, using averages of ultracompact radio sources, the angular-size versus redshift relation has been measured out to 2 ~ 3 and the minimum in the curve verified.

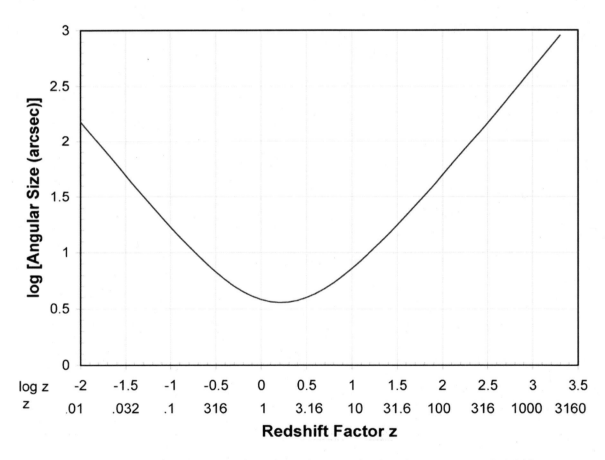

Fig. 17.3-1. Apparent Angular Size of a Standard Galaxy versus Redshift.

Questions

17-1. What was the temperature of the matter which emitted the cosmic background radiation? Why does the radiation appear to have a temperature of 2.73 K today?

17-2. How does the brightness of the cosmic background radiation compare with the brightnesses of other celestial sources?

17-3. What is the connection between non-uniformities in the cosmic background radiation and the large-scale structure of the universe which we see today?

17-4. How do we know that the expansion rate of the universe is accelerating?

17-5. What causes this acceleration?

17-6. Explain why the expansion rate at first showed deceleration and then acceleration as the universe evolved.

17-7. What are the four distances used in cosmology?

17-8. Which distance is the best measure of the apparent distance of an object?

17-9. The universe is 13.7 billion years old. Do we actually see any objects which look 13.7 billion light years away? Explain.

17-10. What is the difficulty in using the plot of angular-size versus redshift?

Problems

17-1. When the universe was about half its present age, the cosmic background radiation appeared to have a temperature of 5.46 K, i.e. twice the present value. Following Example Problem 17.1-1, calculate the total flux and the combined magnitude.

17-2. Using Eqs. 17.3-1 and 17.3-2, calculate the luminosity distance and the now distance at the turnaround point where $z = 1.645$ and $d_{as} = 5.76$ billion light years.

Laboratory Exercise
LE1: The Moon's Orbit

LE1.1 Instructions

<u>Objective:</u> To determine the characteristics of the Moon's orbit from the daily variation of its apparent diameter during two periods, one near minimum eccentricity and one near maximum eccentricity, and compare them.

<u>Background:</u> Due to the perturbing influence of the Sun, the eccentricity of the Moon's orbit varies in a sinusoidal manner with a period of 206 days. Apogee and perigee do not recur at the same position in successive orbits but advance and retreat, and the angular separation between apogee and perigee departs positively and then negatively from 180°, all with the same periodicity. These departures from an ideal constant elliptical orbit can be easily seen by plotting the orbits at the two extreme positions in the cycle and comparing them.

<u>Procedure:</u> In LE1.2, the data for February–March 2004 are the daily values of the Moon's angular diameter near minimum eccentricity and the data for June–July 2004 are the daily values of its angular diameter near maximum eccentricity. Half of the class can do the plot and the calculations for one set of data and the other half can do it for the other. The procedural steps are written for the case of minimum eccentricity (February–March 2004), when there are two apogees and one perigee. The words in parentheses refer to the case of maximum eccentricity (June–July 2004), when there are two perigees and one apogee.

1. For each day, calculate the distance of the Moon, r, in units of 10^3 km from the given angular diameter, α, in units of arc seconds, by using the following relation:

$$r = 717,000 / \alpha$$

2. Plot the distance r versus the longitude ϕ from 0° to the last point before 360° on the polar graph paper provided. Connect the points with a smooth curve, which approximates the Moon's orbit.

3. From the data table find the distances of the two apogees (perigees) and the one perigee (apogee). Average the two apogee (perigee) values to obtain a single value. Let r_A be the apogee distance and r_P the perigee distance.

4. Calculate the eccentricity by the following equation:

$$e = (r_A - r_P) / (r_A + r_P)$$

5. From the data table find the longitudes of the two apogees (perigees) and the one perigee (apogee). If the apogee or perigee is closer to being right at the position given use the given value as the longitude. If it is close to halfway between two dates average the two adjacent longitudes to obtain a single value.

6. Calculate the difference in longitude $\Delta\phi_A$ ($\Delta\phi_P$) for the two apogee (perigee) values:

$$\Delta\phi_A = \phi_{A2} - \phi_{A1} \text{ (for February–March 2004)}$$
$$\Delta\phi_P = \phi_{P2} - \phi_{P1} \text{ (for June–July 2004)}$$

7. Average the two apogee (perigee) longitude values to obtain a single value, ϕ_A (ϕ_P), and apogee-to-perigee (perigee-to-apogee) angle:

$$\phi_P - \phi_A \text{ (for February–March 2004)}$$
$$\phi_A - \phi_P \text{ (for June–July 2004)}$$

Questions

1. What causes the Moon's orbit to depart from an ideal constant elliptical orbit?

In Questions 2–5 use your own values of e_{min} (e_{max}), $\Delta\phi_A$ ($\Delta\phi_P$) and the other values from the blackboard as tabulated by your instructor.

2. By what percentage does the eccentricity vary compared to its average value e = .055? Calculate:

$$[(e_{max} - e) / e] \times 100\%$$
$$[(e_{min} - e) / e] \times 100\%$$

3. Does apogee or perigee show the greater difference (advance) in longitude during the orbit?

4. Which orbit has an apogee-to-perigee or perigee-to-apogee angle which is closer to the ideal value of 180°?

5. During which orbit would you expect the Moon to produce the highest tides on the Earth?

LE 1.2 Angular Diameter of the Moon versus Longitude for Two Periods in 2004

Date	Longitude (°)	Angular Diameter (″)	Date	Longitude (°)	Angular Diameter (″)
26 Feb 04	0.0	1791	1 June 04	0.0	1974
27 Feb 04	12.0	1780	2 June 04	14.9	1994
28 Feb 04	23.9	1774	3 June 04	30.1	2005
29 Feb 04	35.7	1775	4 June 04	45.3	2006
1 Mar 04	47.6	1781	5 June 04	60.5	1996
2 Mar 04	59.6	1792	6 June 04	75.4	1977
3 Mar 04	71.8	1807	7 June 04	90.0	1951
4 Mar 04	84.3	1826	8 June 04	104.2	1922
5 Mar 04	97.1	1847	9 June 04	117.9	1892
6 Mar 04	110.2	1869	10 June 04	131.1	1864
7 Mar 04	123.6	1889	11 June 04	144.0	1838
8 Mar 04	137.3	1907	12 June 04	156.6	1817
9 Mar 04	151.0	1922	13 June 04	168.9	1799
10 Mar 04	165.3	1932	14 June 04	180.9	1785
11 Mar 04	179.5	1938	15 June 04	192.9	1775
12 Mar 04	193.7	1941	16 June 04	204.8	1768
13 Mar 04	207.9	1939	17 June 04	216.7	1764
14 Mar 04	222.1	1936	18 June 04	228.5	1764
15 Mar 04	236.2	1930	19 June 04	240.3	1766
16 Mar 04	250.2	1921	20 June 04	252.2	1772
17 Mar 04	264.2	1911	21 June 04	264.2	1780
18 Mar 04	277.8	1898	22 June 04	276.2	1793
19 Mar 04	291.3	1884	23 June 04	288.5	1809
20 Mar 04	304.7	1868	24 June 04	300.9	1828
21 Mar 04	317.8	1850	25 June 04	313.7	1852
22 Mar 04	330.7	1832	26 June 04	326.8	1879
23 Mar 04	343.3	1814	27 June 04	340.3	1908
24 Mar 04	355.7	1799	28 June 04	354.2	1937
25 Mar 04	7.9	1786	29 June 04	8.5	1964
26 Mar 04	19.9	1777	30 June 04	23.3	1986
27 Mar 04	31.7	1773	1 Jul 04	38.3	2001

28 Mar 04	43.6	1774	2 Jul 04	53.6	2006
29 Mar 04	55.4	1781	3 Jul 04	68.8	2000
30 Mar 04	67.4	1793	4 Jul 04	83.8	1984

Laboratory Exercise
LE2: Orbits of Visual Binary Stars

LE2.1 Instructions

<u>Objective:</u> To determine the true orbits of two visual binary stars from their observed projected orbits and to determine the individual component masses using Kepler's third law and given values of the mass ratio.

<u>Background:</u> The apparent orbit of either component of a visual binary star around the other one is an ellipse, but its major and minor axes are not the major and minor axes of the true orbit. However, by the use of a parallel-ruler device, the true major and minor axes may be drawn on the apparent orbit, yielding the true eccentricity and axial ratio. From the apparent axial ratio, the true axial ratio and the major-to-minor axis angle, the rotation and tilt angles of the apparent orbit may be determined from an interpolation table and the true semimajor axis calculated. Given the period of the orbit, the sum of the masses may be calculated by Kepler's third law and, given the ratio of masses, the individual masses determined.

<u>Procedure:</u> The apparent orbit of a nearby visual binary star, namely α CMa (Sirius) is shown in LE2.2. Consult Fig. 12.5-1.
1. Place the parallel ruler device on the apparent orbit and pin it through the hole in one of the cross bars to the focus F_1 so that it is free to pivot about F_1.
2. Pivot the device about F_1 and adjust the parallel rulers until you find an orientation where the parallel rulers are tangent to the apparent ellipse on both sides.
3. Make a mark through the hole in the other cross bar onto the paper and label it Point Q. Draw lines following the parallel rulers and tangent to the ellipse on either side.
4. Remove the parallel ruler and draw the true major axis passing through Points F_1 and Q and intersecting the apparent ellipse at Points P (closest to F_1) and A (farthest away). Measure its length PA, divide by two to obtain the apparent semimajor axis a' and mark the center Point C on the axis (CA = CP = a')
5. Draw the minor axis passing through C and intersecting the apparent ellipse at the points of tangency B_1 and B_2. Measure its length B_1B_2 and divide by two to obtain the apparent semiminor axis b'.
6. Calculate the true eccentricity e and true axial ratio b/a:
$$e = CF_1/CP = CF_1/CA$$
$$b/a = \sqrt{(1 - e^2)}$$

7. Calculate the apparent axial ratio b'/a' and the axial ratio quotient:
$$q = (b'/a') / (b/a)$$

8. Measure the apparent major-to-minor axis angle δ', as measured from the CP portion of the major axis to the minor axis with a protractor.

9. Use the interpolation table to determine the rotation angle ϕ and tilt angle θ from your values of q and δ'. If δ' lies between 90° and 180°, use 180° − δ'. Find a cell in the interpolation table which contains a pair of values (q, δ') close to your own. For several rows above and below this cell estimate where your value of q lies and draw a curve connecting the points. Do the same for δ'. From the point of intersection of these curves estimate the values of ϕ and θ to the nearest degree.

10. Calculate the true semimajor axis a:

$$a = a'\sqrt{[(1 + q^2)/(1 + \cos^2\theta)]}$$

Convert a to arc sec using the scale 1 arc sec = 8.77 mm. Divide a in mm by 8.77 to obtain a in arc sec.

11. Convert a to AU using a distance of 2.637 parsecs

Multiply a in arc sec by 2.637 to obtain a in AU.

$$a\ (AU) = a\ (\text{arc sec}) \times d\ (\text{parsecs})$$

12. Find the sum of the masses using Kepler's third law:

$$M_A + M_B = a^3/P^2$$

where P = 50.09 yr.

13. Find the individual masses given the mass ratio determined from the relative velocities as M_B/M_A = .47. Substitute M_B = .47 M_A in the above equation and solve for M_A. Then solve for M_B.

Questions

1. Why cannot the apparent major axis a′ be used in Kepler's third law to determine the sum of the masses?
2. What device did you use and what did you do with it to determine the true major axis?
3. What is the major-to-minor axis angle in the true orbit, i.e. not rotated and tilted? Look at Fig. 2.5–1.
4. What law did you use to determine the sum of the masses?
5. What additional information did you have to be given to determine the individual masses?

LE2.2 Apparent Orbit of a Visual Binary Star

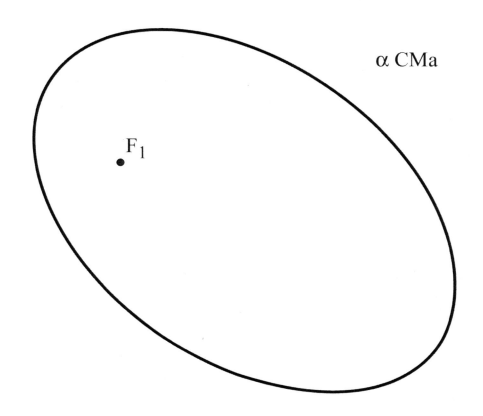

Laboratory Exercise
LE3: The Blackbody Spectrum and Stellar Spectra

LE3.1 Instructions

<u>Objective:</u> To determine the color temperatures and the goodness of fit for a set of stars spanning a range of spectral classes by fitting blackbody spectra to their photometric spectra.

<u>Background:</u> The blackbody spectrum is the ideal spectrum emitted by a hot opaque object at a uniform temperature T. Stars like the Sun and those somewhat cooler have spectra fairly close to the blackbody ideal but significant departures from the ideal occur at either end of the spectral sequence. For the very cool late-M giant stars, large absorption bands due to molecules such as TiO_2 make the spectrum depart from a smooth blackbody spectrum. In the A and late-B stars, a feature known as the Balmer jump, which is a sudden drop of flux in the ultraviolet, occurs. It is produced by the ionization of hydrogen from its first excited state, and gives rise to a two-peaked, distinctly non-blackbody spectrum. In this laboratory exercise we will fit stellar photometric spectra both by eye and by a computer program to determine the goodness of fit.

<u>Procedure:</u> The values of the normalized flux for seven stars spanning a range of spectral classes are given in the table in LE3.2. Plots of these spectra for all the stars except R Dor are shown in Figs. 11.1-1, 12.3-1b, 12.2-2a, 12.3-2b, 12.3-3a, and 12.3-3b.

1. For the six stars having plots of the spectra place the transparency of the blackbody spectrum over the stellar spectrum and slide it back and forth to achieve a best fit by eye. Note where the log λ = 0 position on the transparency overlays the log λ scale on the stellar spectrum. This is log λ_{max}.
2. Determine the color temperature of the stellar spectrum by using the following relations:

$$\log T = 6.565 - \log \lambda_{max}$$
$$T = 10^{\wedge} \log T$$

3. Execute the computer program BBFIT. For each star, enter your value of T from Step 2 and calculate the difference area A. Vary T in either direction until a minimum value of the difference area is attained. (For R Dor use an initial value for T of 2000 K.)
4. Summarize your results in the following table:

Star	Spectrum	Fit by Eye		Fit with Computer Program	
		$\log \lambda_{max}$	T	T	A
η Uma	B3V				
δ Per	B5III				
α Lyr	A0Va				
α Aql	A7V				
Sun	G2V				
α Ori	M1.5Ia/ab				
R Dor	M8IIIq:e				

Questions

1. Which of the stars has the closest fit by a blackbody spectrum, i.e. minimum difference area?
2. Which one has the worst fit?
3. What is the Balmer jump and what causes it?
4. What causes a departure from the blackbody ideal in the very cool late-M giant stars?
5. Based on your results, what value of the minimum difference area would you expect a K1 star to have? (Hint: K1 is approximately midway between G2 and M1.5)

LE3.2 Photometric Spectra of Selected Stars

Filter	log λ	Normalized Flux for						
		η Uma B3V	δ Per B5III	α Lyr A0V	α Aql A7V	Sun G2V	α Ori M1.5 Ia/ab	R Dor M8IIIq:e
m110	2.041	1.523	.672	.065	0	0	0	0
m133	2.124	1.934	1.274	.352	.009	0	0	0
m155	2.190	1.940	1.554	.850	.047	0	0	0
m191	2.281	1.441	1.285	.840	.338	0	0	0
m246	2.391	.967	.965	.629	.302	.036	0	0
m298	2.474	.771	.903	.762	.538	.536	0	0
U	2.544	.616	.936	1.036	1.017	.912	.007	0
B	2.643	.743	1.215	2.318	2.455	1.693	.104	.003
V	2.740	.396	.691	1.469	1.906	1.695	.375	.008
R	2.845	.206	.408	.823	1.261	1.492	.924	.055
I	2.954	.106	.226	.490	.868	1.180	1.837	.615
J	3.097	.052	.104	.280	.515	.800	1.590	1.231
H	3.217	.024	.057	.141	.287	.507	1.566	2.332
K	3.336	.011	.021	.065	.131	.260	.964	1.711
L	3.549	.003	.007	.017	.037	.090	.361	.721
M	3.681	.001	.002	.008	.016	.023	.129	.314

Laboratory Exercise
LE4: Our Place in the Milky Way Galaxy

LE4.1 Instructions

Objective: To determine the distances and three-dimensional positions of two globular clusters from the apparent brightness of their RR Lyrae variables and to determine the position of the Sun in the Milky Way galaxy from these positions and those of 55 additional globular clusters.

Background: Because of the extinction of starlight by dust, the distribution of visible stars in the plane of the disk cannot be used to locate the Sun's position because most of the galaxy remains invisible. Objects outside the disk plane visible at large distances must be used and the best such objects are the globular clusters, of which there are ~160 as satellites of the Milky Way. A majority of the older globular clusters contain RR Lyrae variable stars, which are pulsating stars having periods in the range .4 day - .6 day and amplitudes of .5 magnitude to 1.5 magnitudes. Their average absolute magnitude is 0.75, i.e. they are 40-50 times brighter than the Sun. If the overall average magnitude of the RR Lyrae variables in a cluster is measured, and corrected (magnitude value reduced) for interstellar extinction by measuring the reddening of the spectrum, then an accurate distance to the cluster can be calculated and its three dimensional position (x,y,z) with respect to the Sun determined. Here x is the distance toward or away from the center of the galaxy, y the distance toward or away from the direction of rotation and z the distance above or below the galactic plane. The older globular clusters have a roughly spherical distribution about the center of the Milky Way galaxy so the average x-position of a number of globular clusters, <x>, is the distance from the Sun to the center.

Procedure: The maximum and minimum magnitudes of the RR Lyrae variables for two globular clusters, namely NGC 6121 and NGC 6981, are given in LE4.2 and the (x, y, z) coordinates for 55 other globular clusters are given in LE4.3

1. Find the average magnitude m_V for all of the RR Lyrae variables for clusters NGC 6121 and NGC 6981 using $m_V = (m_{V_{max}} + m_{V_{min}})/2$
2. Find the overall average $<m_V>$, the standard deviation $\sigma(m_V)$ and the uncertainty $\sigma(m_V)/\sqrt{N}$, where N is the number of RR Lyrae variables in the set, for both clusters.
3. Calculate the corrected average magnitude $<m_V>'$ using the values of Δm_V given.
4. Calculate the logarithm of the distance and the distance d to the clusters.
5. Calculate the three dimensional positions (x, y, z) of the clusters using the given values of the galactic longitude and galactic latitude.

6. Enter the (x, y, z) values in the blank spaces for these clusters in the table in LE4.3, Note that the coordinates in the table are given in kiloparsecs (10^3 pc).
Convert your values of x, y and z to kiloparsecs before entering them in the table.
7. Find the averages and the standard deviations of the three coordinates for all 57 globular clusters in the table. The average of the x-coordinate, <x> gives the approximate distance of the center of the Milky Way galaxy from the Sun.

Questions

1. Why do we use globular clusters to determine the position of the Sun in the Milky Way?
2. What stars are useful as distance indicators to the globular clusters? What are their periods and amplitude ranges?
3. How do the offsets in the other two directions, i.e. <y> and <z> compare with the offset in the x-direction? Calculate <y>/<x> and <z>/<x> as fractions.
4. If the distribution of the globular clusters is truly spherical the standard deviations $\sigma(x)$, which are measures of the spread of the cluster distribution in the three directions, should be approximately equal. Let us say they should all be within 25% of the average of the three standard deviations $(\sigma(x) + \sigma(y) + \sigma(z)) / 3$. Is this so?
5. RR Lyrae variables come from solar type stars which have evolved to the stage of helium burning. Why don't the younger globular clusters contain RR Lyrae variables?

LE4.2: Maximum and Minimum Magnitudes of RR Lyrae Variables in Two Globular Clusters

NGC 6121 (M4)				NGC 6981 (M72)		
$\phi = -8.10°$, $\theta = 15.8°$, $\Delta m_v = -1.40$				$\phi = 35.0°$, $\theta = -32.8°$, $\Delta m_v = -0.06$		
m_{vmax}	m_{vmin}	$\overline{m}_v$		m_{vmax}	m_{vmin}	$\overline{m}_v$
13.88	14.56			16.45	17.25	
14.20	14.57			16.29	17.95	
13.46	14.66			16.16	17.74	
13.14	14.18			16.56	17.74	
13.89	14.28			16.40	17.43	
13.10	14.66			16.36	17.53	
13.14	14.69			16.73	17.74	
13.44	14.70			16.73	17.54	
13.20	14.52			16.69	17.77	
13.55	14.65			16.81	17.72	
13.30	14.76			16.31	17.17	
13.36	14.78			15.77	16.85	
13.27	14.69			16.40	17.06	
13.54	14.48			16.63	17.56	
12.84	14.20			16.31	17.21	
13.16	14.62			16.57	17.62	
12.73	14.10			15.70	16.28	
13.40	13.98			16.50	17.40	
13.97	14.72			16.56	17.86	
13.65	14.53			16.95	17.73	
13.08	14.51			16.20	16.55	
13.14	14.43			16.30	17.78	
13.10	14.26			16.83	17.64	
13.13	14.27			16.68	17.48	
13.24	14.20			16.44	17.36	
13.11	14.02			16.84	17.78	

12.92	14.04			16.78	17.75	
13.32	14.73					
13.61	14.44					
13.21	13.81					
Average: $<\overline{m}_v>$				Average: $<\overline{m}_v>$		
Standard Dev: $\sigma(\overline{m}_v)$				Standard Dev: $\sigma(\overline{m}_v)$		
Uncertainty $= \sigma(\overline{m}_v)/\sqrt{N}$				Uncertainty $= \sigma(\overline{m}_v)/\sqrt{N}$		
$<\overline{m}_v>' = <\overline{m}_v> + \Delta m_v$				$<\overline{m}_v>' = <\overline{m}_v> + \Delta m_v$		
$\log d = (<\overline{m}_v>' - .75)/5 + 1$				$\log d = (<\overline{m}_v>' - .75)/5 + 1$		
Distance: $d = 10^{\log d}$	pc			Distance: $d = 10^{\log d}$	pc	
$x = d \cos\theta \cos\phi$	pc			$x = d \cos\theta \cos\phi$	pc	
$y = d \cos\theta \sin\phi$	pc			$y = d \cos\theta \sin\phi$	pc	
$z = d \sin\theta$	pc			$z = d \sin\theta$	pc	

LE4.3: Coordinates for Selected Globular Clusters of the Milky Way

	Coordinates (10^3 pc)				Coordinates (10^3 pc)		
	x	y	z		x	y	z
NGC 6441	11.5	−1.3	−1.0	NGC 6715 (M5A)	25.9	2.5	−6.5
NGC 6304	6.0	−0.4	0.6	NGC 4499	10.7	−14.0	−6.6
NGC 6388	9.6	−2.5	−1.2	NGC 5139 (wCen)	3.2	−3.9	1.4
NGC 6569	10.6	0.1	−1.2	NGC 7089 (MZ)	5.6	7.5	−6.7
NGC 6362	5.9	−4.1	−2.3	NGC 7008	17.3	35.1	−13.8
NGC 6652	9.9	0.3	−2.0	NGC 6656 (M22)	3.1	0.5	−0.4
NGC 6638	9.4	1.3	−1.2	RUP 106	10.7	−17.8	4.3
NGC 6712	6.3	3.0	−0.5	NGC 5286	7.2	−8.1	2.0
NGC 6171 (M107)	5.9	0.3	2.5	NGC 6139	9.6	−3.0	1.2
NGC 6723	8.3	0.0	−2.6	PAL 13	1.0	18.9	−17.5
NGC 362	3.1	−5.0	−6.2	NGC 6093 (M80)	9.4	−1.2	3.3
NGC 6864 (M75)	17.5	6.5	−9.0	NGC 6333 (M9)	7.7	0.7	1.5
NGC 6121 (M4)				NGC 4833	3.6	−5.4	−0.9
NGC 1951	−4.3	−8.9	−6.9	NGC 5897	10.2	−3.1	6.2
NGC 5904 (M5)	5.1	0.3	5.4	NGC 6809 (M55)	4.8	0.7	1.5
NGC 6266 (M62)	6.8	−0.8	0.9	NGC 4147	3.6	−5.4	−0.9
NGC 6284	15.1	−0.4	2.6	NGC 2298	10.2	−3.1	6.2
NGC 1261	0.1	−10.1	−12.9	NGC 5824	26.3	−13.7	12.0
NGC 6642	8.3	1.4	−0.9	NGC 5634	15.7	−5.0	19.1
NGC 6402 (M14)	8.4	3.3	2.4	NGC 6293	8.7	−0.4	1.2
NGC 6981 (M72)				NGC 5024 (M53)	2.8	−1.4	17.5
NGC 6229	6.5	22.2	19.7	NGC 4590 (M68)	4.1	−7.1	6.0
NGC 6522	7.7	0.1	−0.5	NGC 5466	3.3	3.0	15.2
NGC 6626 (M28)	5.5	0.8	−0.5	NGC 6426	17.6	9.4	5.8
NGC 6584	12.3	−4.0	−3.8	NGC 7078 (M15)	3.9	8.3	−4.7
NGC 6205 (M13)	3.0	5.0	5.0	NGC 6341 (M92)	2.5	6.3	4.7
NGC 6934	9.1	11.7	−5.1	NGC 5053	2.9	−1.3	16.1
NGC 5272 (M3)	1.5	1.4	10.2				

NGC 3201	0.6	−4.9	0.8				
NGC 5986	9.3	−4.0	2.4				
		Average \<x\>, \<y\>, \<z\> Std. Dev: σ(x), σ(y), σ(z)					

Distance to the Center of the Milky Way = \<x\> = _____ $\times 10^3$ pc

= _____ $\times 10^3$ ly

Laboratory Exercise
LE5: Hubble's Law and High Redshifts

LE5.1 Instructions

<u>Objective:</u> To verify Hubble's law and determine Hubble's constant using Type Ia supernovae as standard candles and to measure the redshifts and calculate the recession velocities of two quasars with high redshifts.

<u>Background:</u> Because of the expansion of the universe more distant objects have higher redshifts. The relation is approximately linear for recession velocities small compared with the speed of light: $v = Hd$, where H is Hubble's constant. Type Ia supernovae are useful as standard candles because they are very bright and fairly uniform in brightness, since they all come from the same type of star, namely a white dwarf whose mass is right at the Chandrasekhar limit of 1.38 solar masses. From the inverse-square law, we can assume that the distances of the supernovae are inversely proportional to the square root of their brightnesses. The recession velocities can be determined from the redshift of the Si II line in the red and infrared part of the spectrum. A plot of recession velocity v vs. distance d yields a linear relation whose slope is the Hubble constant H. Quasars, which are the nuclei of distant active galaxies, are much brighter than Type Ia supernovae and can be seen to much greater distances, but are not useful as standard candles because of their large range of intrinsic brightness and their variability.

<u>Procedure:</u> The spectrum of one Type Ia supernovae, namely SN2006bq, is given in Fig. 13.3-1a. The spectra of four others are given in LE5.2. The spectra of two quasars with high redshifts are given in Figs. 16.2-2a and 16.2-2b. Use the following table for the supernova data.

Supernova	Flux @ log λ = 2.625	Flux rel. to SN2006bq	Rel. Dist. = $1\sqrt{\text{(Rel. Flux)}}$	Distance d (Mpc)	log λ	Δlog λ	Velocity v (km/s)
SN2006bq		1.00	1.00	188			
SN2006bw							
SN2006ak							
SN2002cc							
SN2001bp							

1. Estimate the flux values in the blue part of the spectra at log λ = 2.625 and write them in Column 1.
2. Calculate the relative flux values, i.e. the flux divided by the flux for SN2006bq and write them in Column 2.
3. Calculate the relative distances as one over the square root of the relative fluxes and write them in Column 3.

4. Multiply the relative distances by the distance for SN2006bq (188 Mpc) and write them in Column 4.
5. Measure the shift of the minimum of the Si II absorption line in the red portion of the spectrum, $\Delta\log \lambda$, from an assumed rest position of $\log \lambda = 2.785$ ($\lambda \sim 610$ nm) and write them in Column 5.
6. Calculate the recession velocities using the relation $v = 2.3 \times c \times \Delta\log \lambda$ and write them in Column 6.
7. Plot recession velocity vs. distance d on the graph paper provided. Draw a best-fit straight line which passes through the origin and has minimum overall error with the points. Measure the slope of the line which is the Hubble constant H.
8. Place the transparency of the composite quasar spectrum over each of the quasar spectra and slide it back and forth until the positions of the spectral lines coincide. Determine the spectral shift $\Delta\log \lambda$ and calculate the redshift z and the ratio v/c from the following equations:

$$z = 10^{\Delta\log \lambda} - 1$$
$$v/c = [(1 + z)^2 - 1]/[(1 + z)^2 + 1]$$

Questions

1. Why are Type Ia supernovae useful as standard candles?
2. Why are quasars not useful as standard candles?
3. What law did you use to determine the relative distances of the supernovae from their relative brightnesses?
4. Why must the line on the v vs. d plot go through the origin?
5. How did you determine the Hubble constant H from the v vs. d plot?

LE5.2 Spectra of Four Type Ia Supernovae

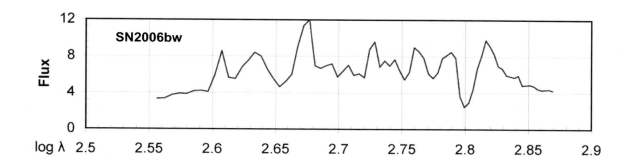

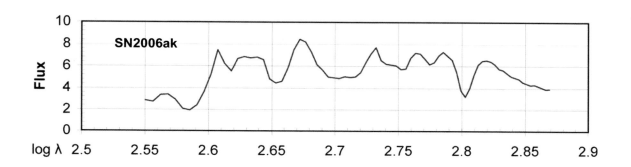

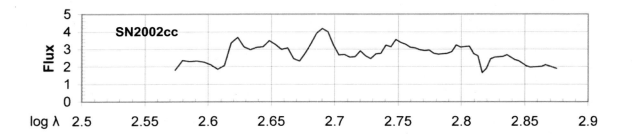

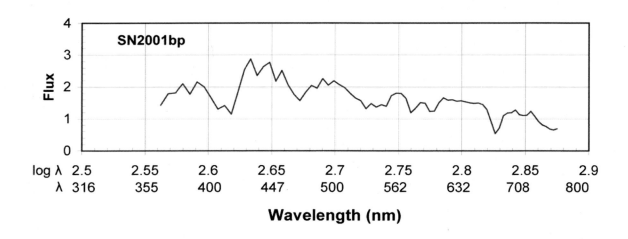

Wavelength (nm)

Appendix A1

Fundamental Physical Constants		
Small Constants		
Planck's constant h	6.62608×10^{-34} Js	(joule-seconds)
Mass of electron m_e	9.1094×10^{-31} kg	(kilograms)
Mass of proton m_p	1.673×10^{-26} kg	(kilograms)
Charge of electron e	1.609×10^{-19} C	(coulombs)
Gravitational constant G	$6.67259 \times 10^{-11} m^3/kg/s^2$	(meters cubed per kilogram per second squared)
Boltzmann's constant k	1.380×10^{-23} J/K	(joules per Kelvin)
Stefan-Boltzmann Constant σ	5.6705×10^{-8} W/m^2/k	(watts per meter squared per Kelvin fourth)
Large Constants		
Speed of Light c	2.997925×10^8 m/s $\approx 3.00 \times 10^8$ m/s	(meters per second)

Some Important Astronomical Quantities	
Distances	
Astronomical Unit (AU)	$= 1.495979 \times 10^{11}$ m $= 1.50 \times 10^8$ km
Light-year (ly)	$= 9.4605 \times 10^{15}$ m $= 6.324 \times 10^4$ AU
Parsec (pc)	$= 3.085678 \times 10^{16}$ m $= 2.06265 \times 10^5$ AU
Sizes	
Radius (equatorial) of the Earth	$= 6.378 \times 10^6$ m $= 6.378 \times 10^3$ km
Radius (polar) of the Earth	$= 6.357 \times 10^6$ m $= 6.357 \times 10^3$ km
Radius of the Sun (R_o)	$= 6.96 \times 10^8$ m $= 6.96 \times 10^5$ km
Times	
Year (sidereal) (y)	$= 3.155815 \times 10^7$ s $= 365.256366$ d
Masses	
Mass of the Earth	$= 5.974 \times 10^{24}$ kg
Mass of the Moon	$= 7.349 \times 10^{22}$ kg
Mass of the Sun (M_o)	$= 1.989 \times 10^{30}$ kg
Luminosities	
Luminosity of the Sun (L_o)	$= 3.83 \times 10^{26}$ W

Appendix A2: Useful Equations for Elliptical Orbits

1. Equation of ellipse in polar coordinates:

$$r = a\,(1 - e^2) / [1 + e\cos(\psi - \psi_p)]$$

where r = distance from focus F_1
a = semimajor axis
e = eccentricity
ψ = angle around the orbit
ψ_p = angle of perihelion/perigee/etc.

2. Distances at perihelion/perigee/etc. and aphelion/apogee/etc.:

$$r_P = a\,(1 - e)$$
$$r_A = a\,(1 + e)$$
$$e = (r_A - r_P) / (r_A + r_P)$$

3. Axial ratio related to eccentricity:

$$b/a = \sqrt{(1 - e^2)}$$
where b = semiminor axis

4. Elliptical orbit rotated by angle ϕ (between $-90°$ and $90°$) and tilted by angle θ (between $0°$ and $90°$).

Let (a) a and b be the semimajor and semiminor axes of the true orbit
(b) a' and b' be the semimajor and semiminor axes of the projected orbit
(c) $q = (b'/a') / (b/a)$
(d) δ' = angle between the projected major and minor axes

Then q and δ' are given in terms of ϕ and θ by:

$$q = \sqrt{[(\sin^2\phi\cos^2\theta + \cos^2\phi) / (\cos^2\phi\cos^2\theta + \sin^2\phi)]}$$
$$\cos\delta' = (1 + q)\,[\sin\phi\cos\phi\,(1 - \cos^2\theta) / (\cos^2\phi\cos^2\theta + \sin^2\phi)]$$

These formulas cannot be inverted to obtain ϕ and θ from q and δ', but the values of q and δ' may be tabulated and ϕ and θ determined by interpolation. Also, the true major axis is given by

$$a/a' = \sqrt{[(1 + q^2) / (1 + \cos^2\theta)]}$$

Appendix A3: Interpolation Graph for Projected Elliptical Orbits

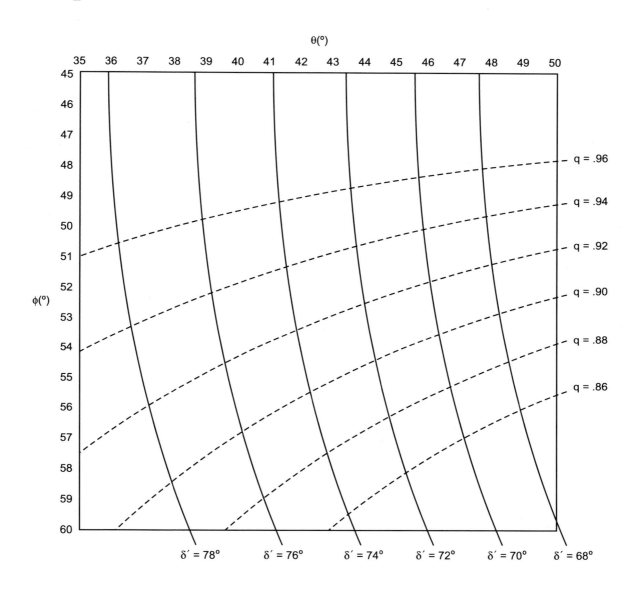

Appendix A4: Procedure for Fitting a Blackbody Spectrum to a Stellar Photometric Spectrum

Some sample photometric spectra of stars are given in the table in LE3.2. The spectra are normalized so that the total area under the curve is unity. The fitting procedure is as follows: Given an input value of the temperature T, a blackbody spectrum can be calculated at the same values of wavelength at which the photometric spectrum was measured, a <u>difference area</u> between the two curves calculated, and the temperature T varied to achieve an minimum difference area. The equations used to compute the normalized flux f_{bb} of the blackbody spectrum are as follows:

$$\lambda_i = 10^{\wedge}\log \lambda_i$$
$$x_i = hc \: / \: \lambda_i \: kT$$
$$f_{bbi} = (15 \times 10 \: / \: \pi^4) \: x_i^4 \: / \: (e^{xi} - 1)$$

where i is the index on spectral points and runs from 1 to 16. The difference between the stellar spectrum and the blackbody spectrum at the ith point is

$$\Delta f_i = f_i - f_{bbi}$$

The difference area for the interval between the ith point and the $(i + 1)$st point depends on the relative sign of Δf_i and Δf_{i+1}. If they have the same sign the difference area is

$$A_i = (\log\lambda_{i+1} - \log\lambda_i) \: (\Delta f_i + \Delta f_{i+1}) \: / \: 2$$

Whereas if they have opposite sign it is

$$A_i = (\log\lambda_{i+1} - \log \lambda_i) \: [\Delta f_i \: / \: (1 - \Delta f_{i+1} \: / \: \Delta f) + \Delta f_{i+1} \: / \: (1 - \Delta f_i \: / \: \Delta f_{i+1})] \: / \: 2$$

The A_i are then summed from 1 to 15 to compute the total difference area of A.

For O and B stars, it is necessary to add some extra points at short wavelengths, e.g. log λ = 1.65, 1.75, 1.85, 1.95, at which the stellar flux is zero because of interstellar hydrogen absorption, but the blackbody flux is appreciable.

Appendix A5: Constellation Names, Abbreviations and Their Meaning

Constellation	Abbr.	Meaning of Name	Constellation	Abbr.	Meaning of Name
Andromeda	And	The Princess of Ethiopia	Coma Berenices*	Com	Berenice's Hair
Antlia	Ant	The Air Pump	Corona Australis	CrA	The Southern Crown
Apus*	Aps	The Bird of Paradise	Corona Borealis	CrB	The Northern Crown
Aquarius	Aqr	The Water Bearer	Corvus	Crv	The Crow
Aquila	Aql	The Eagle	Crater	Crt	The Cup
Ara	Ara	The Altar	Crux	Cru	The Southern Cross
Aries*	Ari	The Ram	Cygnus	Cyg	The Swan
Auriga	Aur	The Charioteer	Delphinus	Del	The Dolphin
Boötes	Boo	The Herdsman	Dorado*	Dor	The Swordfish
Caelum	Cae	The Sculptor's Chisel	Draco	Dra	The Dragon
Camelopardus*	Cam	The Giraffe	Equuleus	Equ	The Little Horse
Cancer*	Cnc	The Crab	Eridanus	Eri	The River
Canes Ventici*	CVn	The Hunting Dogs	Fornax	For	The Furnace
Canis Major	CMa	The Big Dog	Gemini	Gem	The Twins
Canis Minor	CMi	The Little Dog	Grus*	Gru	The Crane
Capricornus	Cap	The Sea Goat	Hercules	Her	Hercules
Carina	Car	The Keel (of Argo)	Horologium	Hor	The Clock
Cassiopeia	Cas	The Queen of Ethiopia	Hydra	Hya	The Water Snake (f)
Centaurus	Cen	The Centaur	Hydrus	Hyi	The Water Snake (m)
Cepheus	Cep	The King of Ethiopia	Indus	Ind	The American Indian
Cetus	Cet	The Whale	Lacerta	Lac	The Lizard
Chameleon*	Cha	The Chameleon	Leo	Leo	The Lion
Circinus	Cir	The Compasses	Leo Minor	LMi	The Little Lion
Columba	Col	The Dove	Lepus*	Lep	The Hare

Constellation	Abbr.	Meaning of Name	Constellation	Abbr.	Meaning of Name
Libra	Lib	The Scales	Pyxis*	Pyx	The Compass (of Argo)
Lupus	Lup	The Wolf	Reticulum	Ret	The Net
Lynx	Lyn	The Lynx	Sagitta	Sge	The Arrow
Lyra	Lyr	The Lyre	Sagittarius	Sgr	The Archer
Mensa	Men	Table (Mountain)	Scorpius	Sco	The Scorpion
Microscopium	Mic	The Microscope	Sculptor	Scl	The Sculptor
Monoceros	Mon	The Unicorn	Scutum	Sct	The Shield
Musca	Mus	The Fly	Serpens	Ser	The Serpent
Norma	Nor	The Square	Sextans	Sex	The Sextant
Octans	Oct	The Octant	Taurus	Tau	The Bull
Ophiuchus	Oph	The Serpent Bearer	Telescopium	Tel	The Telescope
Orion	Ori	The Hunter	Triangulum	Tri	The Triangle
Pavo	Pav	The Peacock	Triangulum Australe	TrA	The Southern Triangle
Pegasus	Peg	The Winged Horse	Tucana	Tuc	The Toucan
Perseus	Per	Perseus	Ursa Major	UMa	The Big Bear
Phoenix	Phe	The Phoenix	Ursa Minor	UMi	The Little Bear
Pictor	Pic	The Easel	Vela*	Vel	The Sails (of Argo)
Pisces*	Psc	The Fishes	Virgo*	Vir	The Virgin
Piscis Austrinus	PsA	The Southern Fish	Volans	Vol	The Flying Fish
Puppis	Pup	The Stern (of Argo)	Vulpecula	Vul	The Fox

Regular Genitive Forms		*Irregular Genitive Forms	
-a → -ae	-o → -onis	Apus → Apodis	Dorado → Doradus
-ans → -antis	-on → -onis	Aries → Arietis	Grus → Gruis
-e → -is	-or → -oris	Camelopardus → Camelopardalis	Lepus → Leporis
-ens → -entis	-os → -otis	Cancer → Cancri	Pisces → Piscium
-er → -eris	-um → -i	Canes Venatici → Canum Venaticorum	Pyxis → Pyxidis
-es → -is	-us → -i	Chameleon → Chameleontis	Vela → Velorum
-i → -orum	-x → -cis	Coma Berenices → Comae Berenices	Virgo → Virginis
-is → -is			

Appendix A6: Star Names, Their Meaning and Origin

Const.	Star	Star Name	Derivation	Meaning	Origin
And	α	Alpheratz	Al Surrat al Faras	The Horse's Navel	Arabic
And	β	Mirach	Al Mizar	The Girdle	Arabic
And	γ	Almach	Al Anak al Ard	a small mammal	Arabic
Aql	α	Altair	Al Nasr al Tair	The Flying Eagle	Arabic
Aql	γ	Tarazed	Shahin Tara Zed	Star-Striking Falcon	Persian
Aql	ζ	Deneb el Okab	Al Danab al Uqab	The Tail of the Eagle	Arabic
Ari	α	Hamal	Al Ras al Hamal	The Head of the Sheep	Arabic
Aur	α	Capella	Capella	Little She-Goat	Latin
Aur	β	Menkalinan	Al Mankib dhil Inan	The Shoulder of the Rein Holder	Arabic
Aur	η	Hoedus II	Hoedus II	Second Kid	Latin
Aur	ι	Hassaleh		East End of the Belt	Arabic
Boo	α	Arcturus	Arktouros	Bear Guard	Greek
Boo	ε	Izar	Al Izar	The Girdle	Arabic
CMa	α	Sirius	Seirios	Sparkling, Scorching	Greek
CMa	β	Murzim	Al Murzim	The Announcer	Arabic
CMa	δ	Wezen	Al Wazn	The Weight	Arabic
CMa	ε	Adhara	Al Adhara	The Virgins	Arabic
CMa	ζ	Furud	Al Furud	The Bright Single Ones	Arabic
CMa	η	Aludra	Al Adhra	The Virgin	Arabic
CMi	α	Procyon	Prokyon	Before the Dog	Greek
CMi	β	Gomeisa	Al Ghumaisa	The Weeping One	Arabic
Car	α	Canopus	Kahi Nub	Golden Earth	Coptic
Car	β	Miaplacidus	Miah Placidus	Quiet Waters	Arabic and Latin
Cas	α	Shedir	Al Sadr	The Breast	Arabic
Cas	β	Caph	Al Kaff al Hadib	The Hand Stained with Henna	Arabic
Cen	α	Rigel Kentaurus	Al Rijl al Kentaurus	The Centaur's Foot	Arabic and Latin
Cen	β	Agena Hadar	Gena	Knee	Latin
Cen	θ	Menkent	Menkalinan Kentauros	Centaur's Shoulder	Arabic and Greek

Const.	Star	Star Name	Derivation	Meaning	Origin
Cep	α	Alderamin	Al Dhira al Yamin	The Right Arm	Arabic
Cep	β	Alfirk	Al Firk	The Flock	Arabic
Cet	α	Menkar	Al Minhar	The Nose	Arabic
Cet	β	Deneb Kaitos	Al Dhanab al Kaitos	The Tail of the Whale	Arabic
Cet	o	Mira	Stella Mira	Wonderful Star	Latin
Col	α	Phact	Al-Fakhitah	Ring Dove	Arabic
CrB	α	Alphecca	Al Nair al Fakkah	The Bright One of the Dish	Arabic
Cru	α	Acrux	Alpha Crux		contraction
Cru	β	Bcrux	Beta Crux		contraction
Cru	γ	Gacrux	Gamma Crux		contraction
Crv	ε	Minkar	Al Minkar al Ghurab	The Raven's Beak	Arabic
Crv	γ	Gienah	Al-Janah	The Wing	Arabic
Cyg	α	Deneb	Al Dhanab al Dajujah	The Hen's Tail	Arabic
Cyg	β	Albireo	Ab Ireo(incorrect transcription of Arabic name)	from the herb ireus(Arabic: The Beak of the Hen)	Latin
Cyg	γ	Sadr	Al Sadr al Dhajujah	The Hen's Breast	Arabic
Cyg	ε	Gienah	Al Junah	The Wing	Arabic
Dra	γ	Etamin	Al Ras al Tinnin	The Dragon's Head	Arabic
Dra	δ	Altais	At Tinnin	The Great Serpent	Arabic
Dra	ζ	Aldhibah	Al Dibah	The Hyenas	Arabic
Eri	α	Achernar	Al Ahir al Nahr	The End of the River	Arabic
Gem	α	Castor	Castor	one of the twins (horseman)	Latin
Gem	β	Pollux	Pollux	one of the twins (pugilist)	Latin
Gem	γ	Alhena	Al Hanah	The Brand or Mark	Arabic
Gem	ε	Mebsuta		Outstretched Paw	Arabic
Gem	η	Propus	Propus	in front of (Castor's left foot)	Latin
Gem	μ	Tejat Posterior		Back Foot	Arabic & Latin
Gru	α	Al Nair	Al Nair	The Bright One	Arabic

Const.	Star	Star Name	Derivation	Meaning	Origin
Her	α	Ras Algethi	Al Ras al Jathiyy	The Kneeler's Head	Arabic
Her	β	Kornephoros	Kornephoros	Club Bearer	Greek
Hya	α	Alphard	Al Fard al Shuja	The Solitary One in the Serpent	Arabic
Leo	α	Regulus	Regulus	diminutive of Rex (king)	Latin
Leo	β	Denebola	Al Dhanab al Asad	The Lion's Tail	Arabic
Leo	γ	Algieba	Al Juba	The Mane	Arabic
Leo	δ	Duhr	Al Thabr al Asad	The Lion's Back	Arabic
Lib	β	Kiffa Borealis	Kiffa Borealis	Northern Scale Tray	Arabic and Latin
Lib	σ	Brachium	Brachium	Arm	Latin
Lyr	α	Vega	Al Nasr al Waki	The Eagle of the Desert	Arabic
Oph	α	Ras Alhague	Al Ras al Hawwe	The Head of the Serpent Charmer	Arabic
Oph	β	Cobalrai	Al Kalb al Rai	The Heart of the Shepherd	Arabic
Oph	δ	Yed Prior	Yed Prior	Former (star) in the Hand	Arabic and Latin
Oph	ζ	Han	Han		Chinese
Oph	η	Sabik	Sabik	precedes	Arabic
Ori	α	Betelgeuse	Al Ibt al Jauzah	The Armpit of the Central One	Arabic
Ori	β	Rigel	Al Rijl Jauzah al Yusra	The Left Leg of the Central One	Arabic
Ori	γ	Bellatrix	Bellatrix	Female Warrior	Latin
Ori	δ	Mintaka	Al Mintakah	The Belt	Arabic
Ori	ε	Alnilam	Al Nitham	The String of Pearls	Arabic
Ori	ζ	Alnitak	Al Nitak	The Girdle	Arabic
Ori	η	Saif al Jabbar	Al Saif al Jabbar	The Sword of the Hunter	Arabic
Ori	κ	Saiph	Al Saif	The Sword	Arabic
Ori	λ	Meissa	Al Maisan	The Proudly Marching One	Arabic
Ori	ι	Nair al Saif	Al Nair al Saif	The Bright One of the Sword	Arabic
Ori	υ	Thabit	Al Thabit	The Imperterbable One	Arabic
Pav	α	Peacock	Peacock		English

Const.	Star	Star Name	Derivation	Meaning	Origin
Peg	α	Markab	Al Markab	The Saddle	Arabic
Peg	β	Scheat	Al Saq	The Leg or Shin	Arabic
Peg	γ	Algenib	Al Janb	The Side	Arabic
Peg	ε	Enif	Al Anf	The Nose	Arabic
Peg	ζ	Hamam	Al Saad al Humam	The Lucky Star of the High-Minded Man	Arabic
Per	α	Mirfak	Al Marfik	The Elbow	Arabic
Per	β	Algol	Al Ras al Ghul	The Demon's Head	Arabic
Per	ζ	Menkib	Al Mankib	The Shoulder	Arabic
Per	ρ	Gorgonea Tertia	Gorgonea Tertia	Third Star of the Gorgon	Latin
Phe	α	Nair al Zaurak	Al Nair al Zaurak	The Bright One of the Boat	Arabic
PsA	α	Fomalhaut	Al Fum al Hut	The Mouth of the Fish	Arabic
Pup	ζ	Suhail Hadr	Al Suhail al Hadr	The Suhail of the Ground	Arabic
Sgr	δ	Kaus Meridionalis	Kaus Meridionalis	Middle of the Bow	Arabic and Latin
Sgr	ε	Kaus Australis	Kaus Australis	Southern Part of the Bow	Arabic and Latin
Sgr	ζ	Ascella	Axilla	Armpit	Latin
Sgr	η	Ain al Rami	Al Ain al Rami	The Eye of the Archer	Arabic
Sgr	λ	Kaus Borealis	Kaus Borealis	Northern Part of the Bow	Arabic and Latin
Sgr	σ	Nunki	Nunki	Of Enki (the Sumerian God of Waters)	Sumerian
Sco	α	Antares	Antares	Similar to Mars	Greek
Sco	β	Graphias	Grapsaios	Crab	Greek
Sco	ε	Wei	Wei	(named for Giange Wei)	Chinese
Sco	θ	Girtab	Girtab	Scorpion	Sumerian
Sco	λ	Shaula	Al Shaulah	The Sting	Arabic
Sco	σ	Alniyat	Al Niyat	The Arteries	Arabic
Sco	υ	Lesath	Al Lasah	The Sting	Arabic
Ser	α	Unuk al Hay	Al Unk al Hayyah	The Neck of the Snake	Arabic
Tau	α	Aldebaran	Al Dabaran	The Follower (of the Pleiades)	Arabic
Tau	β	Alnath	Al Natih	The Butting One	Arabic
Tau	η	Alcyone	Alcyone	one of the Pleiades sisters	Greek

Const.	Star	Star Name	Derivation	Meaning	Origin
UMa	α	Dubhe	Al Tharab al Dubb al Akbar	The Back of the Greater Bear	Arabic
UMa	β	Mirak	Al Maraqq	The Loin	Arabic
UMa	γ	Phecda	Al Falidh	The Thigh	Arabic
UMa	ε	Alioth	Al Jawn	The Black Horse	Arabic
UMa	ζ	Mizar	Al Mizar	The Girdle	Arabic
UMa	η	Alkaid	Al Kaid Burat al Naash	The Leader of the Mourning Maidens	Arabic
UMa	υ	Tania Australis	Ath-Thaniyah	The Southern one of the Second Leap	Arabic & Latin
UMi	α	Polaris	Stella Polaris	Pole Star	Latin
UMi	β	Kochab	Al Kuukab al Shamaliyy	The Star of the North	Arabic
Vel	γ²	Suhail Muhlif	Al Suhail al Muhlif	The Suhail of the Oath	Arabic
Vel	λ	Suhail Wazn	Al Suhail al Wazn	The Suhail of the Weight	Arabic
Vir	α	Spica	Spica	Ear of Wheat	Latin
Vir	δ	Auva	Al Awwa	The Barker	Arabic
Vir	ε	Vindemiatrix	Vindemiatrix	Gatherer of Grapes	Latin

Appendix A7: The Twenty Nearest Star Systems

Star System		Parallax (arc sec)	Distance (ly)	Precision	m_v	M_v	Spectral Type	Period (yr)	a (AU)	e
Sun		-	-	-	-26.72	4.85	G2V			
α Cen	A	.74212	4.395	.002	-0.01	4.34	G2V	79.92	23.69	.516
α Cen	B	.74212	4.395	.002	1.35	5.70	K0V			
α Cen	C	.77233	4.223	.003	11.01	15.45	M5.5V			
Barnard's Star		.54901	5.941	.003	9.54	13.21	M4V			
Wolf 359		.418	7.80	-	13.45	16.6	M5V			
Lalande 21185		.39240	8.312	.002	7.49	10.46	M2V			
Luyten 726-8	A	.381	8.57	.09	12.41	15.3	M5.5V	26.5	5.45	.615
	B				13.2	16.1	M6V			
α CMa (Sirius)	A	.37921	8.601	.004	-1.44	1.45	A1V	50.09	19.78	.592
	B						A2VII			
Ross 154		.33648	9.693	.005	10.37	13.00	M3.5V			
Ross 248		.316	10.33	.04	12.29	14.8	M5.5V			
ε Eri		.31075	10.496	.003	3.72	6.18	K2V			
Lacaille 9352		.30390	10.732	.003	7.35	9.76	M1.5Ve			
Ross 128		.29958	10.887	.007	11.12	13.50	M4V			
Luyten 789-6	A	.294	11.08	.09	12.33	14.7	M5V	3.8d	1.22	.437
	C						M	2.25		
	B						M			
61 Cyg	A	.28713	11.359	.005	5.20	7.49	K5V	653.34	84.66	.400
	B	.28542	11.427	.003	6.05	8.33	K7V			

α CMi (Procyon) A	.28593	11.407	.003	0.40	2.68	F5IV/V	40.82	5.0	.4
B						AVII			
Struve 2398 A	.28448	11.465	.018	9.70	11.97	M3V	408.	56.	.53
B	.28028	11.637	.009	8.94	11.18	M3.5V			
Groombridge 34 A	.28027	11.638	.004	8.09	10.33	M1.5V	2600.	147.	.00
B						M3.5V			
ε Indi	.27576	11.828	.003	4.69	6.89	K5Ve			
GJ 1111	.275	11.83	-	14.79	17.0	M6.5V			
τ Cet	.27417	11.896	.003	3.49	5.68	G8Vp			

Stars by Spectral Class

Spectral Class	Number of Stars	
	Primary	Secondary
AV	1	
FV	1	
GV	3	
KV	3	2
MV	12	6
AVII		2
Total	20	10

Binaries by Period Range

Period Range (yr)	Number of Binaries
.01–.1	1
.1–1	0
1–10	1
10–100	4
100–1,000	2
1,000–10,000	1
Total	9

Appendix A8: Stars With a Combined Magnitude Less Than 2.5

| Star | Spec. Type | m$_v$ | Flux (nW/m²) | | | | m | Galactic | | Mult. Star Form |
			UV	VIS	IR	Total		Long.	Lat.	
γ Vel	WC8+O2.5e+ B1IV	1.78	52.74	3.15	0.64	56.53	−0.89	262.90	−7.69	** * °
ζ Pup	O5Iaf	2.25	38.77	2.04	0.34	41.15	−0.54	255.98	−4.71	*
ξ Per	O7e	4.04	2.13	0.37	0.15	2.65	2.44	160.37	−13.11	*°
S Mon	O7Ve	4.66	3.95	0.22	0.04	4.21	1.93	202.94	2.20	* ° °
λ Ori	O8e+B0.5V	3.39	8.80	.71	.15	9.66	1.03	195.05	−12.00	* ° °
τ CMa	O9Ib	4.40	3.12	.27	.05	3.44	2.15	238.18	−5.54	** * °
ι Ori	O9III	2.77	19.59	1.27	.21	21.07	0.19	209.52	−19.58	* °
10 Lac	O9V	4.88	2.51	.18	.03	2.72	2.41	96.65	−16.98	* °
ζ Ori	O9.5Ibe+B0III	1.76	39.27	3.17	.57	43.01	−0.59	206.45	−16.59	** * °
μ Col	O9.5V	5.17	2.73	.14	.02	2.89	2.34	237.29	−27.10	*
σ Ori	O9.5V+B0.5V	3.81	8.42	.49	.08	8.99	1.11	206.82	−17.34	** *
ζ Oph	O9.5Vn	2.56	5.64	1.41	.39	7.44	1.32	6.28	23.59	*
ε Ori	B0Iae	1.70	34.11	3.36	.64	38.11	−0.46	205.21	−17.24	* °
φ¹ Ori	B0II	4.41	2.75	0.27	0.05	3.07	2.28	195.40	−12.29	*°
δ Ori	B0III+O9V	2.23	26.44	2.06	.35	28.85	−0.16	203.86	−17.74	** °
γ Cas	B0IVe	2.47	19.92	1.65	0.44	22.01	.14	123.58	−2.15	* °
υ Ori	B0V	4.62	4.77	0.25	0.04	5.06	1.73	210.44	−20.98	*
τ Sco	B0V	2.82	22.49	1.23	.21	23.93	0.05	351.54	12.81	*
θ Car	B0Vp	2.76	21.99	1.27	.21	23.47	0.07	289.60	−4.90	*

Star	Spec. Type	m_v	UV	VIS	IR	Total	m	Long.	Lat.	Mult. Star Form
			\<-- Flux (nW/m²) --\>					\<-- Galactic --\>		
κ Ori	B0.5Iav	2.06	20.67	2.56	.45	23.68	0.06	214.51	-18.50	*
J Pup	B0.5Ib	4.24	2.71	.32	.07	3.10	2.27	262.06	-10.42	*
β Cru	B0.5III	1.25	74.79	5.13	0.81	80.73	-1.27	302.46	3.18	** °
λ Lep	B0.5IV	4.29	5.45	.32	.05	5.82	1.58	214.83	-26.24	*
α Cru	B0.5IV+B1V	.79	120.37	7.97	1.19	129.53	-1.79	300.13	-0.36	* *
ξ¹ CMa	B0.5IV	4.33	4.14	0.30	0.05	4.49	1.86	232.11	-14.53	* °
δ Sco	B0.5IV	2.32	15.36	1.83	.39	17.58	0.38	350.10	22.49	** ° °
ε Per	B0.5V	2.89	11.88	1.11	.20	13.19	0.69	157.35	-10.09	* °
1886–7 Ori	B0.5V+BIV	4.38					(1.7)	209.56	-19.71	* *
ρ Leo	B1Ib	3.85	2.80	.45	.09	3.34	2.19	234.89	52.77	**
γ Ara	B1Ib	3.34	5.29	.72	.14	6.15	1.52	334.64	-11.48	* °
ζ Per	B1Ib	2.85	2.45	1.84	.36	3.85	2.03	162.29	-16.69	* ° °
β CMa	B1II/III	1.98	34.22	2.65	0.42	37.29	-0.43	226.06	-14.27	* °
ε Cen	B1III	2.30	24.16	1.93	.31	26.40	-0.06	310.19	8.72	* °
β Cen	B1III	0.61	116.08	9.10	1.36	126.54	-1.76	311.77	1.25	*° °
α Vir	B1III/IV+B2V	0.97	87.13	6.67	1.10	94.90	-1.45	316.11	50.84	** °
β Cep	B1IV	3.23	10.51	0.83	0.14	11.48	0.84	107.54	14.03	* ° °
ω¹ Sco	B1V	3.96	2.61	.39	.10	3.10	2.27	352.75	22.77	*
η Ori	B1V+B2e	3.35	8.14	0.72	0.13	8.99	1.11	204.87	-20.39	** *
β Sco	B1V+B2V	2.62	11.92	1.36	.30	13.58	0.66	353.19	23.60	** ° * *
π Sco	B1V+B2V	2.89	12.37	1.10	.19	13.66	0.66	347.22	20.23	** °
42 Ori	B1V	4.59					(2.4)	208.50	-19.11	* °
α Lup	B1.5III	2.30	18.09	1.93	.31	20.33	0.22	321.61	11.44	* °
κ Sco	B1.5III	2.41	19.94	1.75	0.28	21.97	0.14	351.04	-4.72	*
κ CMa	B1.5IVne	3.96	4.57	.43	.11	5.11	1.72	242.36	-14.99	*

Star	Spec. Type	m_v	Flux (nW/m²)					Galactic		Mult. Star Form
			UV	VIS	IR	Total	m	Long.	Lat.	
δ Lup	B1.5IV	3.22	9.09	.85	.22	10.16	0.98	331.32	13.82	*
μ¹ Sco	B1.5IV	3.08	9.45	0.94	0.15	10.54	0.94	346.12	3.91	*°
η Cen	B1.5Vne	2.31	15.56	1.90	0.33	17.79	0.37	322.77	16.67	*
ε CMa	B2II	1.50	34.29	4.05	.69	39.03	−0.48	239.83	−11.33	* °
ε Cas	B2III	3.38					(1.8)	129.84	1.65	*
γ Ori	B2III	1.64	32.41	3.58	0.60	36.59	−0.41	196.93	−15.95	* °
β Lup	B2III	2.68	14.18	1.37	.24	15.79	0.50	326.25	13.91	*
ν Eri	B2III	3.93					(1.8)	199.31	−31.38	* °
σ Lup	B2III	4.42					(2.4)	318.93	9.25	*
σ Sco	B2III+O9.5V	2.89	4.63	1.37	.69	6.69	1.43	351.31	17.00	** *
π⁴ Ori	B2III+B2IV	3.69					(2.0)	192.89	−23.52	**
N Sco	B2III/IV	4.23	2.48	0.32	0.06	2.86	2.35	345.94	9.22	*
κ Cen	B2IV	3.13					(1.1)	326.87	14.75	* °
Φ Cen	B2IV	3.83	3.72	0.47	0.08	4.27	1.92	315.98	19.07	*
λ Sco	B2IV	1.63	44.05	3.63	.63	48.31	−0.72	351.74	−2.21	*° °
δ Cet	B2IV	4.07	3.52	.38	.06	3.96	2.00	170.76	−52.21	**
ζ Cas	B2IV	3.66	4.69	.55	.09	5.33	1.68	120.78	−8.91	*
γ Peg	B2IV	2.83	12.66	1.19	.20	14.05	0.63	109.43	−46.68	*° °
μ² Sco	B2IV	3.57	6.08	0.60	0.10	6.78	1.42	346.20	3.86	*
ν Cen	B2IV	3.41	8.01	0.70	0.12	8.83	1.13	314.41	19.89	*°
δ Cen	B2IVne	2.60	11.93	1.35	.32	13.60	0.66	296.00	11.57	* *
γ Lup	B2IV	2.78	11.75	1.24	.21	13.20	0.69	333.19	11.89	* *
θ Oph	B2IV	3.27	8.65	.81	.14	9.60	1.04	0.46	6.65	**
δ Cru	B2IV	2.80					(0.6)	298.23	3.79	*

Star	Spec. Type	m_v	Flux (nW/m²)					Galactic		Mult. Star Form
			UV	VIS	IR	Total	m	Long.	Lat.	
υ Sco	B2IV	2.69	14.71	1.38	.23	16.32	0.46	351.27	−1.84	*
α Pav	B2IV	1.94	20.19	2.70	.58	23.47	0.07	340.91	−35.19	*° °
ρ Sco	B2IV/V	3.86	3.99	0.46	0.08	4.53	1.85	344.63	18.27	* °
υ¹ Cen	B2IV/V	3.87	4.22	0.45	0.08	4.75	1.80	315.29	16.45	*
ε Lup	B2IV/V	3.37	5.47	0.73	0.14	6.34	1.99	329.23	10.32	*° °
κ Vel	B2IV/V	2.50					(0.6)	275.88	−3.54	*°
α Mus	B2IV/V	2.69					(0.7)	301.66	−6.30	* °
ε Lup	B2IV/V	3.37					(0.5)	329.23	10.32	*° *
a Car	B2IV/V	3.44					(0.4)	277.69	−7.37	*°
μ¹ Cru	B2IV/V	4.03					(2.1)	303.36	5.69	* *
ω CMa	B2IV/Ve	3.85	2.30	.48	.10	2.88	2.35	239.41	−7.15	*
μ Cen	B2IV/Ve	3.04	6.61	0.98	0.19	7.78	1.27	314.24	19.12	* °
λ Eri	B2IVne	4.27					(2.3)	209.14	−26.69	*
σ Cen	B2V	3.91					(1.9)	299.10	12.47	*
χ Cen	B2V	4.36	2.62	.29	.06	2.97	2.31	317.73	19.54	*
Φ Per	B2Vep	4.07	2.69	.36	.12	3.17	2.24	131.32	−11.33	**
α Ara	B2Vne	2.95	6.94	1.04	.23	8.21	1.21	340.76	−8.83	* °
β Mus	B2V+B3V	3.05					(1.2)	302.45	−5.24	* *
ζ Cen	B2.5IV	2.55	19.22	1.54	0.25	21.01	0.19	314.07	14.09	*°
η Lup	B2.5IV	3.41	7.22	0.69	0.11	8.02	1.23	338.77	11.01	* °
ι Lup	B2.5IV	3.55	4.01	0.60	0.11	4.72	1.81	318.48	14.14	*
τ Lib	B2.5V	3.66					(2.0)	341.06	20.45	*
ζ Cru	B2.5V	4.04					(2.3)	299.33	−1.36	* °
σ Sgr	B2.5V	2.02					(−0.2)	9.56	−12.43	* °
ζ CMa	B2.5V	3.02	6.64	0.98	0.16	7.78	1.27	237.52	−19.43	*° °

Star	Spec. Type	m_v	Flux (nW/m²)					Galactic		Mult. Star Form
			UV	VIS	IR	Total	m	Long.	Lat.	
o² CMa	B3Ia	3.02	3.27	.95	.22	4.44	1.88	235.55	−8.23	*
δ Pic	B3III+O9V	4.81	2.69	.22	.04	2.95	2.32	263.30	−27.68	** °
ι Her	B3IV	3.80	2.90	.48	.09	3.47	2.14	72.32	31.27	** °
α Tel	B3IV	3.51					(1.9)	348.67	−15.18	*
o Vel	B3IV	3.62					(2.0)	270.25	−6.80	*°
χ Car	B3IVp	3.47					(1.9)	266.68	−12.32	*
η Aur	B3V	3.17	5.13	0.86	0.16	6.15	1.52	165.35	0.27	*
ρ Cen	B3V	3.96					(2.4)	296.78	10.03	*
5471 Cen	B3V	4.06	2.81	.40	.06	3.27	2.21	325.90	20.10	*
λ Lup	B3V	4.05	2.63	.38	.07	3.08	2.27	326.80	11.13	* *
η UMa	B3V	1.86	17.94	2.87	.53	21.34	0.17	100.69	65.32	*
α Eri	B3Vp	0.46	48.11	10.29	2.11	60.51	−0.96	290.84	−58.79	*
λ Tau	B3V+A4IV	3.47					(2.3)	178.37	−29.38	*°°
β Mon	B3Ve+B3ne+B3e	3.74	3.04	0.5	.11	3.65	2.09	216.66	−8.21	** * * °
ζ Tau	B4IIIpe	3.00	6.13	1.00	0.23	7.36	1.33	185.69	−5.64	**
PP Car	B4Vne	3.32	3.32	.74	.19	4.25	1.92	287.18	−3.15	*
δ Per	B5III	3.01	3.04	.98	.22	4.24	1.93	150.28	−5.77	* °
η CMa	B5Ia	2.45	3.91	1.65	.41	5.97	1.55	242.62	−6.49	*
ζ Dra	B6III	3.17	2.00	.83	.18	3.01	2.30	96.01	35.04	*
β Tau	B7III	1.65	10.31	3.43	0.73	14.47	0.59	177.99	−3.74	* °
η Tau	B7IIIe	2.87	2.07	1.10	.28	3.45	2.15	166.67	−23.45	** ° ° °
α Col	B7IVe	2.64	3.43	1.38	.32	5.13	1.72	238.81	−28.86	* °
α Gru	B7IV	1.74	9.98	3.15	.67	13.80	0.65	350.00	−52.47	* °

Star	Spec. Type	m$_v$	Flux (nW/m^2)				m	Galactic		Mult. Star Form
			UV	VIS	IR	Total		Long.	Lat.	
α Leo	B7V	1.35	9.40	4.48	1.00	14.88	0.56	226.43	48.94	* °
β Ori	B8Iae	0.12	23.75	13.30	3.43	40.48	−0.52	209.24	−25.25	* ° ° °
Φ Sgr	B8III	3.17	1.74	.85	.21	2.80	2.38	8.00	−10.77	**
ν Pup	B8III	3.17	1.64	0.84	0.19	2.67	2.43	251.94	−20.54	*
Y Cor	B8IIIp	2.59					(1.9)	290.99	44.50	*
β Lib	B8V	2.61					(1.9)	352.02	39.23	*
β CMi	B8Ve	2.90	2.02	1.07	.26	3.35	2.18	209.52	11.68	* °
β Per	B8V	2.12	4.51	2.16	0.57	7.24	1.35	148.98	−14.90	*° °
α And	B9V	2.06	4.99	2.36	0.51	7.86	1.26	111.73	−32.84	*° °
α Peg	B9V	2.49	1.48	1.53	0.40	3.41	2.16	88.28	−40.38	*
ε Sgr	B9.5III	1.85	2.40	2.73	.76	5.89	1.57	359.20	−9.81	* °
ε UMa	A0p	1.77					(1.5)	122.18	61.16	*
θ Aur	A0p	2.62					(2.3)	179.34	6.73	* °
δ Gem	A0IV	1.92	1.63	2.55	.72	4.90	1.77	196.77	4.45	*° °
α Lyr	A0Va	0.03	12.84	14.28	3.86	30.98	−0.23	67.44	19.24	* °
α CrB	A0V	2.23	1.74	1.92	.52	4.18	1.94	41.87	53.77	**°
γ UMa	A0Ve	2.44	1.17	1.56	.43	3.16	2.25	140.84	61.38	*
Y Cen	A1IV	2.17					(1.9)	301.26	13.88	* *
δ Vel	A1V	1.96					(1.7)	272.08	−7.37	* ° ° °
α CMa	A1V	−1.46	50.91	56.15	14.58	121.64	−1.72	227.22	−8.88	* °
β UMa	A1V	2.37	1.27	1.70	.46	3.43	2.16	149.17	54.80	*
α Gem	A1V+A2Vne	1.59	2.93	3.41	.99	7.33	1.33	187.44	28.48	** ** °°
ζ UMa	A1Vp+A1m	2.06					(1,8)	113,11	61.58	** *

Star	Spec. Type	m_v	Flux (nW/m²)					Galactic		Mult. Star Form
			UV	VIS	IR	Total	m	Long.	Lat.	
α Cyg	A2Iae	1.25	2.25	4.58	1.65	8.48	1.17	84.28	2.00	* °
β Aur	A2IV	1.90	1.47	2.57	.74	4.78	1.80	167.46	10.41	*° °
β Car	A2IV	1.68	1.94	3.18	.95	6.07	1.54	285.98	–14.41	*
η Oph	A2V+A3V	2.43					(2.3)	6.72	14.01	** *
α PsA	A3V	1.16	2.40	4.92	1.51	8.83	1.13	20.49	–64.90	*
β Leo	A3V	2.14	1.01	2.01	.63	3.65	2.09	250.65	70.81	* °
α Oph	A5III	2.08	.81	2.11	.79	3.71	2.07	35.90	22.57	*°
α Cep	A7V	2.45	.99	1.49	.60	2.56	2.47	101.00	9.17	* °
α Aql	A7V	0.77	2.51	6.85	2.83	12.19	0.78	47.74	–8.91	** °
l Car	A8Ib	2.25					(2.3)	278.46	–7.01	*
α Car	F0II	–0.72	7.17	28.81	12.09	48.07	–0.71	261.21	–25.29	*
θ Sco	F0II	1.87					(1.9)	347.14	–5.98	*
β Cas	F2III/IV	2.27	.41	1.72	.91	3.04	2.29	117.52	–3.27	* °
α Per	F5Ib	1.79					(1.8)	146.57	–5.86	* °
α CMi	F5IV/V	0.38	2.40	9.74	5.80	17.94	0.36	213.69	13.03	* °
α UMi	F7Ib/IIv	2.02	.48	2.02	1.62	4.15	1.95	123.28	26.46	*° ° ° °
δ CMa	F8Ia	1.84	.29	2.39	1.96	4.59	1.84	238.42	–8.27	*
γ Cyg	F8Ib	2.20	.17	1.71	1.39	3.27	2.21	78.15	1.87	* °
α Cen	G2V+KIV	–.28	2.13	17.06	15.28	34.47	–0.35	315.78	–0.71	* *
α Aur	G5IIIe+G0III	0.08	1.23	12.02	13.34	26.59	–0.07	162.58	4.57	** °°
ε Boo	K0II/III	2.37					(2.0)	39.38	64.78	* °
β Cet	K0III	2.01	.08	2.07	2.85	5.00	1.75	111.31	–80.68	*

| Star | Spec. Type | m$_v$ | Flux (nW/m²) | | | | m | Galactic | | Mult. | Star Form |
			UV	VIS	IR	Total		Long.	Lat.		
α Phe	K0III	2.39					(2.1)	320.02	-73.98	*	
ε Cyg	K0III	2.46	.06	1.33	2.05	3.49	2.15	75.94	-5.71	*	
α UMa	K0IIIa	1.79	.09	2.53	4.18	6.80	1.41	142.85	51.01	*	°
α Cas	K0IIIa	2.23	.05	1.64	2.76	4.45	1.87	121.42	-6.30	**	°
β Gem	K0IIIb	1.14	.19	4.57	6.46	11.22	0.87	192.23	23.41	*	
θ Cen	K0IIIb	2.06					(1.8)	319.46	24.08	**	°
T Pup	K1III	2.93					(2.5)	260.16	-20.86	**	°
γ¹ Leo	K1IIIb	1.98	.07	2.13	4.01	6.21	1.51	216.55	54.65	**	*
λ Sgr	K1IIIb	2.81	.04	.97	1.57	2.58	2.47	7.66	-6.52	*	
α Boo	K1.5III	-0.05	.37	14.41	32.18	46.96	-0.68	15.14	69.11	*	
ε Peg	K2Ib	2.39	.02	1.51	3.88	5.41	1.66	65.57	-31.46	*	
α TrA	K2IIb/IIIa	1.92					(1.3)	321.54	-15.26	*	
α Ari	K2III	2.00	.06	2.08	3.85	5.99	1.55	144.57	-36.20	*	
β Oph	K2III	2.77	.03	1.00	1.75	2.78	2.38	29.21	17.19	*	
ε Crv	K25IIIa	3.00	.02	.85	1.80	2.67	2.43	290.59	39.26	*	
ε Sco	K2.5III	2.29	.04	1.60	2.77	4.41	1.88	348.82	6.56	*	
α Ser	K2IIIb	2.65	.03	1.14	2.02	3.19	2.24	14.20	44.08	*	
π Pup	K3Ib	2.70	.02	1.24	4.69	5.95	1.56	249.01	-11.28	**	°
β Ara	K3Ib/IIa	2.85	.01	.97	2.50	3.48	2.14	335.37	-11.01	*	
β Cyg	K3II+B0.5V+B8Ve	2.95					(2.0)	62.11	4.57	**	*
4050 Car	K3IIa	3.40					(2.5)	285.48	-3.80	**	°
ι Aur	K3II	2.69	.01	1.15	3.48	4.64	1.83	170.59	-6.16	*	
γ Aql	K3II	2.72	.01	1.13	3.22	4.36	1.9	48.73	-7.08	**	°
π Her	K3IIab	3.16	.01	.72	2.03	2.76	2.39	60.66	34.34	*	
γ¹ And	K3IIb	2.10	.06	1.96	4.43	6.45	1.47	136.96	-18.56	*	°

Star	Spec. Type	m_v	Flux (nW/m²)				m	Galactic		Mult.	Star Form
			UV	VIS	IR	Total		Long.	Lat.		
α Hya	K3II/III	1.98	.03	2.23	5.93	8.19	1.21	241.49	29.05	*	°
α Tuc	K3III	2.82	.02	1.04	2.49	3.55	2.12	330.22	−47.96	*°	
ζ Ara	K3III	3.13					(2.4)	332.80	−8.20	*	
δ Sgr	K3IIIa	2.70					(2.0)	3.00	−7.15	*	°
ε Car	K3III+B2V	1.86					(1.2)	274.29	−12.60	**	
β UMi	K4III	2.08	.02	2.08	6.39	8.49	1.17	112.65	40.50	*·	°
λ Vel	K4III	2.21	.02	1.94	7.32	9.28	1.08	265.94	2.82	*°	
ε Lep	K5III	3.19	.01	.75	2.22	2.98	2.31	223.25	−32.73	*	
α Tau	K5III	0.85	.07	6.75	34.95	31.27	−0.24	180.97	−20.25	*°	
γ Dra	K5III	2.23	.02	1.85	6.13	8.00	1.24	79.06	29.22	*°	
σ Pup	K5III	3.25					(2.3)	255.74	−11.91	*°	
σ CMa	K7Ib	3.47					(2.3)	237.17	−10.27	*°	
α Lyn	K7IIIab	3.13	.01	.83	3.16	4.00	1.99	190.24	44.73	*	
μ UMa	M0III	3.05	.01	.92	3.85	4.78	1.80	177.90	56.36	*°	
β And	M0IIIa	2.06	.02	2.26	9.51	11.79	0.82	127.10	−27.10	*	°
γ Phe	M0IIIa	3.41					(2.2)	280.52	−72.17	*°	
δ Oph	M0.5III	2.75	.01	1.22	5.33	6.56	1.45	8.85	32.20	*	°
γ Eri	M0.5III	2.95	.01	1.00	4.16	5.17	1.71	205.16	−44.47	*°	
γ Hyi	M1III	3.26	.01	.78	3.84	4.63	1.83	289.13	−37.8	*	
α Ori	M1.5Ia/Iab	.40	.06	12.12	71.75	83.93	−1.31	199.79	−8.96	*°	°
α Sco	M1.5Iab	.91	.08	7.51	52.51	60.10	−0.95	351.95	15.06	*°	
α Cet	M1.5IIIa	2.53	.01	1.53	7.77	9.31	1.07	173.32	−45.59	*	
μ Cep	M2Iae	4.08	0	.57	7.40	7.97	1.24	100.60	4.32	*°	
CE Tau	M2Ia/Iab	4.38	0	.37	3.48	3.85	2.03	187.18	−8.07	*	
BE Cam	M2IIab	4.47	0	.31	2.73	3.04	2.29	140.02	8.75	*	

Star	Spec. Type	m_v	Flux (nW/m²)					Galactic		Mult.	Star Form
			UV	VIS	IR	Total	m	Long.	Lat.		
β Peg	M2.5II/III	2.42	.01	1.84	12.46	14.31	0.61	95.74	−29.05	*	°
λ Aqr	M2.5IIIa	3.74	0	.50	2.91	3.41	2.16	62.19	−55.73	*	
η Per	M3Ib/IIa	3.76					(1.6)	139.14	−3.18	*	°
π Aur	M3II	4.26	0	.38	3.48	3.86	2.12	166.60	10.94	*	
μ Gem	M3IIab	2.88	.01	1.27	8.84	10.11	.98	189.72	4.17	*	°
η Gem	M3III	3.28	.01	.83	5.98	6.81	1.41	185.85	2.52	**	°
δ Vir	M3III	3.38	.01	.77	5.26	6.03	1.54	305.53	66.25	*	°
YY Psc	M3III	4.41	0	.29	2.46	2.70	2.42	91.58	−65.83	*	
σ Lib	M3IIIa	3.29	.01	.93	6.13	7.06	1.37	337.22	28.62	*	
γ Cru	M3.5III	1.63	.04	4.18	30.49	34.71	−0.36	300.17	5.45	*	°
η Sgr	M3.5III	3.11	.01	1.01	7.75	8.77	1.14	356.44	−9.68	*	°
τ⁴ Eri	M3.5IIIa	3.69	0	.60	4.57	5.17	1.71	212.09	−56.00	*	°
ρ Per	M4II	3.39	.01	.91	9.07	9.99	1.00	149.60	−17.01	*	
δ² Lyr	M4II	4.30	0	.39	4.55	4.94	1.76	66.93	15.32	*	°
54 Eri	M4III	4.32					(2.1)	217.79	−37.36	*	*
ψ Phe	M4III	4.41	0	.34	2.90	3.24	2.22	274.35	−67.22	*	
γ Ret	M4III	4.51	0	.30	2.47	2.77	2.39	274.81	−43.21	*	
62 Sgr	M4III	4.58	0	.32	3.12	3.44	2.15	13.92	−26.93	*	
δ² Gru	M4.5III	4.11	.01	.45	3.70	4.16	1.95	353.28	−57.06	*	
α Her	M5Ib/II+(G5III+A3V)	3.06	.02	1.52	33.68	35.22	−0.37	35.53	27.82	*	**
T Cet	M5IIe	5.12					(1.1)	77.51	−80.20	*	
R Cen	M5IIev	6.39	0	.06	2.76	2.82	2.37	313.42	1.21	*	°
β Gru	M5III	2.10	.03	3.17	35.47	38.67	−0.47	346.27	−57.95	*	
R Lyr	M5III	4.04	.01	.61	10.10	10.71	0.92	73.81	17.79	*	°

Star	Spec. Type	m_v	Flux (nW/m²) UV	VIS	IR	Total	m	Galactic Long.	Lat.	Mult. Star Form
ε Mus	M5III	4.11	0	.53	5.92	6.45	1.47	299.75	-5.30	*
2 Cen	M5III	4.19	0	.55	7.21	7.76	1.27	316.32	26.91	*
RR UMi	M5III	4.60	0	.28	3.84	4.12	1.96	104.87	46.53	*°
ε Oct	M5III	5.10					(2.4)	310.04	-34.50	*
L₂ Pup	M5IIIe	5.10	0	.33	9.60	9.93	1.00	255.77	-15.04	* °
5512 Boo	M5IIIab	5.63	0	.13	2.89	3.02	2.29	14.55	60.79	*
δ Aps	M5IIIb	4.68	0	.41	3.71	4.12	1.96	312.42	-19.89	* *
g Her	M6III	5.06	0	.34	9.38	9.72	1.03	66.16	43.72	*
NU Pav	M6III	5.13	0	.23	5.87	6.10	1.53	337.85	-31.96	*
RX Lep	M6III	5.68	0	.19	5.28	5.47	1.65	212.55	-27.51	*
ρ²Ari	M6III	5.91	0	.14	3.58	3.72	2.07	159.93	-35.45	*
EU Del	M6III	6.25	0	.16	4.05	4.21	1.93	62.05	-13.63	*
RS Cnc	M6IIIase	5.95	0	.19	6.60	6.79	1.42	194.50	42.08	*
U Ori	M6.5III	5.4					(1.1)	188.72	-2.49	*
θ Aps	M6.5III	5.50	0	.21	8.52	8.73	1.14	307.22	-14.54	*
o Cet	M7III	3.05	.02	1.45	18.00	19.47	0.27	167.75	-57.98	*° °
R Hor	M7IIIe	4.0					(0.4)	265.46	-57.38	*
R Cas	M7IIIe	4.80	0	.34	7.64	7.98	1.24	114.56	-10.62	*
R Hya	M7IIIe	4.97	.01	.40	13.79	14.20	0.61	314.22	38.75	* °
R Ser	M7IIIe	5.2					(0.8)	26.23	46.76	*
R Aql	M7IIIe	6.09	0	.11	3.09	3.20	2.23	41.95	0.45	* °
R Leo	M8IIIe	6.02					(0.2)	223.72	44.16	*
R Dor	M8IIIq:e	5.40	0	.27	37.36	37.53	-0.44	272.67	-39.34	* ° °

Star	Spec. Type	m_v	Flux (nW/m²)				m	Galactic		Mult. Star Form
			UV	VIS	IR	Total		Long.	Lat.	
o¹ Ori	S3.5	4.74					(1.0)	185.43	−18.39	*
π¹ Gru	S5	6.62	0	.16	8.90	9.06	1.10	350.28	−55.16	* °
χ Cyg	S6+/1e	4.23	.01	.64	10.22	10.86	0.91	68.54	3.28	* °
U Hya	C5II	4.82	0	.23	2.58	2.81	2.37	259.97	38.07	*
TX Psc	C5II	5.04	0	.20	2.32	2.52	2.49	93.28	−55.60	*
Y CVn	C7I	4.99	0	.21	2.71	2.92	2.33	126.45	71.65	*

Key to Columns in the Preceding Table

Column	Explanation
1	Star designation (4-digit number = number in Yale Bright Star Catalog, 1982)
2	Spectrum taken from Yale Bright Star Catalog (1982)
3	Visual magnitude taken from Yale Bright Star Catalog (1982)
4	Ultraviolet flux (nW/m^2) over the wavelength range 91.15–400 nm.
5	Visible flux (nW/m^2) over the wavelength range 400–800 nm.
6	Infrared flux (nW/m^2) over the wavelength range 800–5000 nm.
7	Total Flux F (nW/m^2) (sum of columns 4, 5, and 6).
8	Combined magnitude M = –2.5 log (F/25).
9	Galactic longitude in degrees.
10	Galactic latitude in degrees.
11	Multiple star form: *: component having $\geq$ 10% of brightness of principal component; °: component having < 10% of brightness of principal component; ** or °°: spectroscopic binary components; * or ° °: visual binary components

The Brightest Stars in Combined Magnitude

Rank	Star	Spectrum	m		Rank	Star	Spectrum	m
1	α Cru	B0.5IV+B1V	−1.79		16	ε CMa	B2II	−0.48
2	β Cen	B1III	−1.76		17	β Gru	M5III	−0.47
3	α CMa	A1Vm	−1.72		18	ε Ori	B0Iae	−0.46
4	α Vir	B1III/IV+B2V	−1.45		19	R Dor	M8IIIq:e	−0.44
5	α Ori	M1.5Ia/Iab	−1.31		20	β CMa	B1II/III	−0.43
6	β Cru	B0.5III	−1.27		21	γ Ori	B2III	−0.41
7	α Eri	B3Vp	−0.96		22	α Her	M5Ib/II+(G5III+A3V)	−0.37
8	α Sco	M1.5Iab	−0.95		23	γ Cru	M3.5III	−0.36
9	γ Vel	WC8+O7.5e+B1IV	−0.89		24	α Cen	G2V+K1V	−0.35
10	α Car	F0II	−0.72		25	α Tau	K5III	−0.24
11	λ Sco	B2IV	−0.72		26	α Lyr	A0Va	−0.23
12	α Boo	K1.5III	−0.68		27	σ Sgr	B2.5V	(−0.2)
13	β Ori	B8Iae	−0.64		28	δ Ori	B0III+O9V	−0.16
14	ζ Ori	O9.5Ibe+B0III	−0.59		29	α Aur	G5IIIe+G0III	−0.07
15	ζ Pup	O5Iaf	−0.54		30	ε Cen	B1III	−0.06

Credits

Chapter	Topic	Source and Website
-	Front Cover	Ultraviolet: Galex http://galex.stsci.edu/GR2/ Visible: John Corban (http://www.spacetelescope.org/projects/fits_liberator/ fitsimages/html/john_corban_14.html) Infrared: 2Mass Atlas Image Gallery (http://www.ipac.caltech.edu/2mass/gallery/images_galaxies.html)
2	Determination of 1 AU	JJMO Mars Parallax Project: www.mccarthyobservatory.org
9	Dwarf Planets	http://en.wikipedia.org/wiki/Dwarf_planet
11	Search for Solar Twins	http://www.mso.anu.edu.au/news/archive/2006/03_mar/ http://www.solstation.com/stars2/18sco.htm
12	Nearest Star Systems	The Hipparcos and Tycho Catalogs: http://www.rssd.esa.int/Hipparcos/catalog.html
12	Photometric Spectra	http://cdsweb.u-strasbg.fr/cats/II.htx
12	Ultraviolet Spectra	OAO2 (OAO-2 Ultraviolet **Spectra**: An Atlas of Stellar **Spectra** OAO-2 http://esavo.esa.int/registry/result.jsp?searchMethod= GetResource&identifier=ivo://gov.nasa.gsfc.nssdc/oao2 NASA NSSDC National Space Science Data Center ultraviolet) IUE Atlas of B-Type Stellar Spectra (Walborn+ 1995) IUE Atlas of O-Type Stellar Spectra (Walborn+ 1985) Copernicus (http://archive.stsci.edu/copernicus/obtaining.html) EURD (http://arxiv.org/abs/astro-ph/0012220v1)
13	RR Lyrae Variables	Christine M. Clement, University of Toronto, http://www.astro.utoronto.ca/~cclement/
13	Supernova Spectra	The CfA Supernova Archive, Harvard University, (http://www.cfa.harvard.edu/supernova//SNarchive.html)
14	Radio Spectrum of Cas A	From 'Atmospheric Window and Sky Brightness' (NRAO Library)
14	X-Ray Spectrum of Cas A	Goddard Space Flight Center (http://heasarc.gsfc.nasa.gov/docs/objects/snrs/)
14	Black Holes	William Robert Johnson (http://johnstonsarchive.net/relativity/bhctable.html)

15	Globular Clusters	Christine M. Clement, University of Toronto, http://www.astro.utoronto.ca/~cclement/ William E. Harris, McMaster University http://www.physics.mcmaster.ca/~harris/mwgc.dat
16	Composite Quasar Spectrum	MAST Composite Quasar Spectrum http://archive.stsci.edu/prepds/composite_quasar/
16	High Redshift Quasar Spectra	The 2dF QSO Redshift Survey (2QZ) http://www.2dfquasar.org/
17	Cosmic Background Radiation Spectrum	Chapter 19: G.F. Smoot (LBNL) and D. Scott, Cosmic Background Radiation http://pdg.lbl.gov/2000/microwaverppbook.pdf
17	Distances in Cosmology	Edward L. (Ned) Wright, Cosmology Tutorial, http://www.astro.ucla.edu/%7Ewright/cosmolog.htm